METHODS IN MOLECULAR BIOLOGY

Series Editor
John M. Walker
School of Life and Medical Sciences
University of Hertfordshire
Hatfield, Hertfordshire, AL10 9AB, UK

For further volumes:
http://www.springer.com/series/7651

Enzyme Stabilization and Immobilization

Methods and Protocols

Second Edition

Edited by

Shelley D. Minteer

*Departments of Chemistry and Department of Materials Science and Engineering,
University of Utah, Salt Lake City, UT, USA*

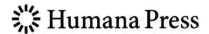 Humana Press

Editor
Shelley D. Minteer
Departments of Chemistry and Materials Science
    and Engineering
University of Utah
Salt Lake City, UT, USA

ISSN 1064-3745   ISSN 1940-6029   (electronic)
Methods in Molecular Biology
ISBN 978-1-4939-8219-6   ISBN 978-1-4939-6499-4   (eBook)
DOI 10.1007/978-1-4939-6499-4

Printed on acid-free paper

This Humana Press imprint is published by Springer Nature
The registered company is Springer Science+Business Media LLC New York

Preface

Enzyme stabilization has been an area of research interest since the 1950s, but in the last three decades, researchers have made tremendous progress in the field. This has opened up new opportunities for enzymes in molecular biology as well as industrial applications, such as bioprocessing, biosensors, and biofuel cells.

The first chapter introduces the reader to the field of enzyme stabilization and the different theories of enzyme stabilization, including the use of immobilization as a stabilization technique. The first part of the book will focus on protocols for enzyme stabilization in solutions including liposome formation, micelle introduction, crosslinking, and additives, while the second part of the book will focus on protocols for enzyme stabilization during enzyme immobilization including common techniques like sol-gel encapsulation, polymer encapsulation, and covalent attachment to supports. Protocols for a variety of enzymes are shown, but the enzymes are chosen as examples to show that these protocols can be used for both enzymes of biological importance and enzymes of industrial importance. The final chapter will detail spectroscopic protocols, methods, and assays for studying the effectiveness of the enzyme stabilization and immobilization strategies.

The chapters of this volume should provide molecular biologists, biochemists, and biomedical and biochemical engineers with the state-of-the-art technical information required to effectively stabilize their enzyme of interest in a variety of environments (i.e., harsh temperature, pH, or solvent conditions).

Salt Lake City, UT, USA   *Shelley D. Minteer*

Contents

Contributors

MIRJANA G. ANTOV • *Faculty of Technology, University of Novi Sad, Novi Sad, Serbia*

SABINA BESIC • *Department of Chemistry, Saint Louis University, St. Louis, MO, USA*

DEJAN I. BEZBRADICA • *Department of Biochemical Engineering and Biotechnology, Faculty of Technology and Metallurgy, University of Belgrade, Belgrade, Serbia*

MICHAEL J. COONEY • *Hawaii Natural Energy Institute, School of Ocean and Earth Science and Technology, University of Hawaii-Manoa, Honolulu, HI, USA*

ALEKSANDRA DIMITRIJEVIĆ • *Department of Biochemistry, Faculty of Chemistry, University of Belgrade, Belgrade, Serbia*

SANJA Ž. GRBAVČIĆ • *Faculty of Technology and Metallurgy, Innovation Centre, University of Belgrade, Belgrade, Serbia*

MUNISHWAR N. GUPTA • *Department of Biochemical Engineering and Biotechnology, Indian Institute of Technology Delhi, New Delhi, India*

AMANDA S. HARPER-LEATHERMAN • *Chemistry and Biochemistry Department, Fairfield University, Fairfield, CT, USA*

DAVID P. HICKEY • *Department of Chemistry, University of Utah, Salt Lake City, UT, USA*

KENISE JEFFERSON • *Department of Chemistry and Biochemistry, Center for Nanoscience, University of Missouri—St. Louis, St. Louis, MO, USA*

GLENN R. JOHNSON • *Hexpoint Technologies, Mexico Beach, FL, USA; Co-located at the Air Force Civil Engineer, Tyndall AFB, FL, USA*

JELENA R. JOVANOVIĆ • *Department of Biochemical Engineering and Biotechnology, Faculty of Technology and Metallurgy, University of Belgrade, Belgrade, Serbia*

JOEL L. KAAR • *Department of Chemical and Biological Engineering, University of Colorado, Boulder, CO, USA*

ZORICA D. KNEŽEVIĆ-JUGOVIĆ • *Department of Biochemical Engineering and Biotechnology, Faculty of Technology and Metallurgy, University of Belgrade, Belgrade, Serbia*

HEATHER R. LUCKARIFT • *Universal Technology Corporation, Dayton, OH, USA; Co-located at the Air Force Civil Engineer, Tyndall AFB, FL, USA*

DUŠAN Ž. MIJIN • *Department for Biochemical Engineering and Biotechnology, Faculty of Technology and Metallurgy, University of Belgrade, Belgrade, Serbia*

NENAD B. MILOSAVIĆ • *Division of Experimental Therapeutics, Department of Medicine, Columbia University, New York, NY, USA*

ROSS D. MILTON • *Department of Chemistry, University of Utah, Salt Lake City, UT, USA*

SHELLEY D. MINTEER • *Departments of Chemistry and Materials Science and Engineering, University of Utah, Salt Lake City, UT, USA*

MICHAEL J. MOEHLENBROCK • *Department of Chemistry, Saint Louis University, St. Louis, MO, USA*

JOYEETA MUKHERJEE • *Chemistry Department, Indian Institute of Technology Delhi, New Delhi, India*

JASMINE M. OLVANY • *Chemistry Department, Lebanon Valley College, Annville, PA, USA*

RADIVOJE M. PRODANOVIĆ • *Department of Biochemistry, Faculty of Chemistry, University of Belgrade, Belgrade, Serbia*

MICHELLE RASMUSSEN • *Chemistry Department, Lebanon Valley College, Annville, PA, USA*

DEBRA R. ROLISON • *U.S. Naval Research Laboratory, Surface Chemistry Branch, Washington, DC, USA*

IPSITA ROY • *Department of Biotechnology, National Institute of Pharmaceutical Education and Research (NIPER), Punjab, India*

JULIA L. RUTHERFORD • *Chemistry Department, Lebanon Valley College, Annville, PA, USA*

OLGA V. SHULGA • *ICL Performance Products, St. Louis, MO, USA*

ANDREA B. STEFANOVIĆ • *Department of Biochemical Engineering and Biotechnology, Faculty of Technology and Metallurgy, University of Belgrade, Belgrade, Serbia*

KEITH J. STINE • *Department of Chemistry and Biochemistry, Center for Nanoscience, University of Missouri—Saint Louis, Saint Louis, MO, USA*

ALICE H. SUROVIEC • *Department of Chemistry and Biochemistry, Berry College, Mt. Berry, GA, USA*

GAIGE R. VANDEZANDE • *Chemistry Department, Lebanon Valley College, Annville, GA, USA*

DUŠAN VELIČKOVIĆ • *Department of Biochemistry, Faculty of Chemistry, University of Belgrade, Belgrade, Serbia*

JEAN MARIE WALLACE • *Nova Research, Inc., Alexandria, VA, USA*

MAKOTO YOSHIMOTO • *Department of Applied Molecular Bioscience, Yamaguchi University, Ube, Japan*

Chapter 1

Introduction to the Field of Enzyme Immobilization and Stabilization

Michael J. Moehlenbrock and Shelley D. Minteer

Abstract

Enzyme stabilization is important for many biomedical or industrial application of enzymes (i.e., cell-free biotransformations and biosensors). In many applications, the goal is to provide extended active lifetime at normal environmental conditions with traditional substrates at low concentrations in buffered solutions. However, as enzymes are used for more and more applications, there is a desire to use them in extreme environmental conditions (i.e., high temperatures), in high substrate concentration or high ionic strength, and in nontraditional solvent systems. This chapter introduces the topic enzyme stabilization and the methods used for enzyme stabilization including enzyme immobilization.

Key words Enzyme immobilization, Enzyme stabilization, Cross-linking, Entrapment, Encapsulation

1 Introduction

This book details methods for enzyme stabilization and enzyme immobilization. Although the concepts of enzyme immobilization and enzyme stabilization are different, they are related. Enzyme immobilization techniques are strategies focused on retaining an enzyme on a surface or support. In theory, they are focused on being able to reuse the enzyme and/or constrain the enzyme to a particular area of a reactor or device, whereas enzyme stabilization is focused on extending the active catalytic lifetime of the protein. Enzyme immobilization is one method for stabilizing enzymes. Enzymes are proteins with complex and fragile three-dimensional structures. The main methods for stabilizing enzymes are: protein engineering, chemical modification, immobilization, and adding additives for stabilization.

Protein engineering is a very common method for enzyme stabilization, but it is out of the scope of this book, because it is not a simple or translational method that can be detailed in a chapter. The common types of protein engineering are directed evolution, site-directed mutagenesis, and peptide chain extensions. Directed

Shelley D. Minteer (ed.), *Enzyme Stabilization and Immobilization: Methods and Protocols*, Methods in Molecular Biology, vol. 1504, DOI 10.1007/978-1-4939-6499-4_1, © Springer Science+Business Media New York 2017

evolution is a common method for enzyme engineering that involves a simple three-step process that can be repeated as many "rounds"/repetitions as desired to evolve the mutant protein to have non-natural attributes (in this case some type of stability). The three-step process involves developing a library of random mutants, testing the library of mutants for the particular property/attribute of interest, selecting the mutant of interest, and then repeating the above steps until optimized. Directed evolution will only be as good as the screening assay that you employ to test the library of mutants for the particular property/attribute of interest. This is frequently done for improving the temperature stability of proteins [1]. Site-direct mutagenesis is another common technique that requires more understanding about the structure of the protein [2]. Rather than random mutations, site-directed mutagenesis involves an examination of the protein structure and intelligently mutating individual sites on the protein structure and then studying the effect of the site mutation on the particular property, activity, or attribute of interest (in this case stability). Peptide chain extensions are not a common technique, in general, for protein or enzyme engineering, but it is a technique for improving stability. Peptide chain extensions are the process of elongating the polypeptide chain of the enzyme on either the N-terminus or C-terminus end. This has been shown to improve temperature stability of proteins [3, 4]. It is not frequently done for biotechnology applications other than improving stability, but it can be effective.

2   Enzyme Stabilization

This book covers the other three methods of enzyme stabilization. The first part of the book focuses on chemical modification and adding additives for stabilization and the second part of the text focuses on enzyme immobilization. Chemical modification is common. The most common method is cross-linking. Cross-linkers are added for intramolecular cross-linking of the peptide chain. This is traditionally done with glutaraldehyde, but other bifunctional reagents can be used, including: dimethyl suberimidate, disuccinimidyl tartarate, and 1-ethyl-3-(3-dimethylaminopropyl)carbodiimide. It is important to note that this is unspecific chemistry as shown in Fig. 1 and therefore can cause both intramolecular and intermolecular cross-linking of the peptide chain, which can cause aggregation. This aggregation can be considered advantageous or disadvantageous depending on the application. In terms of other chemical modification of the protein, conjugation to water-soluble polymers like poly(ethylene glycol) (PEG) has been shown to be effective at stabilizing enzymes [5–8]. This is frequently referred to as PEGylating the enzyme. PEGylating reagents either contain a reactive function

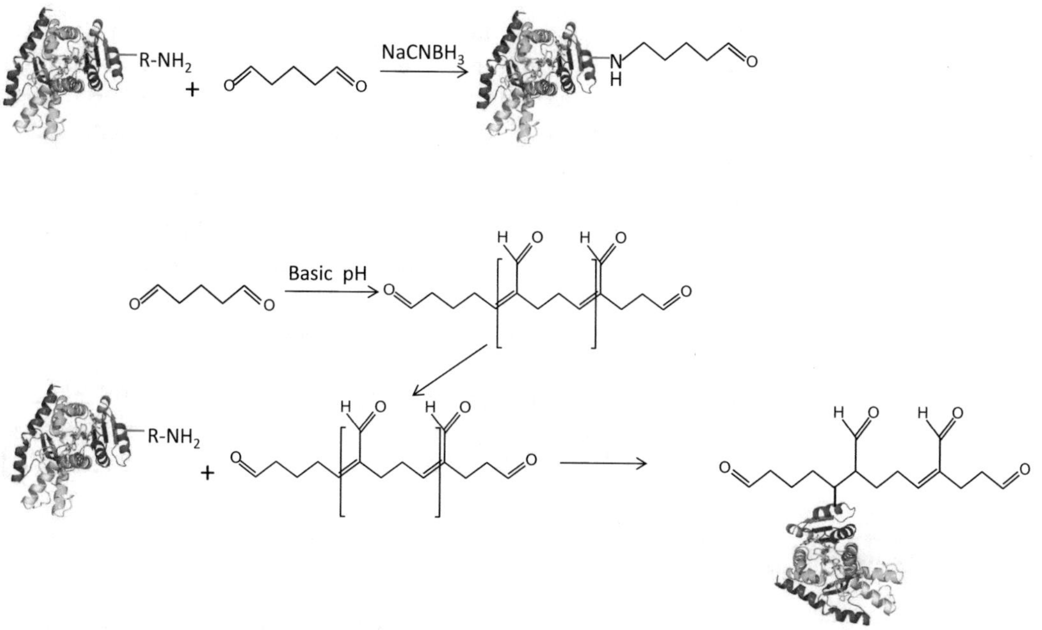

Fig. 1 Options for glutaraldehyde cross-linking chemistry

group at one end of the PEG chain or both ends of the PEG chain (bifunctional). If it is important to reduce intermolecular aggregation, monofunctional PEGylating reagents are typically used with functional groups to either react to the N-terminus or C-terminus end of the protein or to react with individual amino acids, such as aspartic acid, glutamic acid, lysine, tyrosine, cysteine, histidine, or arginine. Other biocompatible polymers can also be employed, such as chitosan [9], alginate, pectin, cellulose, hyaluronate, carrageenan, and poly(glutamate). These are becoming more popular in literature in recent years.

The addition of additives to improve enzyme stabilization is also common. There are four main classes of additives: small molecules, polymer, proteins, and surfactant/micelles. The most common small molecule additive is trehalose. Trehalose is a disaccharide of two glucose units that is used for a variety of biological applications requiring dehydration and rehydration. For this reason, trehalose is a common additive when lyophilizing or freeze drying proteins. The trehalose sugar produces a gel as it dehydrates, thereby protecting proteins, enzymes, organelles, cells, and tissue structure. Many sugars and sugar alcohols have also been employed for improving long term storage and temperature stability, including: glycerol, lactose, mannitol, sorbitol, and sucrose. These are frequently referred to as lyophilization agents, because they improve storage stability during temperature fluctuations associated with freezing and thawing. Finally, polymers and proteins, such as: BSA, human albumin, gelatin, and PEG, have also been employed to improve long-term storage and/or lyophilization. Surfactants are another class of small molecules that are frequently added as additives to increase stability. Proteins have a tendency to denature at interfaces, so surfactants compete for binding at the interface. Also, it has been proposed that surfactants can facilitate refolding of partially denatured proteins. Micelles are also employed. Micelles are particularly useful for membrane-bound proteins, because the amphiphilic nature of the micelle and protein can be matched as well as providing an environment that is much more similar to cellular membranes than the buffer environment [10]. This extends the active lifetime of the protein as well as being beneficial for stabilizing the protein in extreme temperature environments, harsh pH environments, and organic solvents [11, 12].

3 Enzyme Immobilization

Enzyme immobilization is a much broader topic that is arguably part science and part art [13]. There are a number of factors to consider when choosing an enzyme immobilization strategy including: enzyme tolerance to immobilization chemical and physical environment, surface functional groups on the protein, size of the enzyme, charge of the protein/pI, polarity of the protein (hydrophobic/hydrophilic regions), and substrate/product transport needs [14]. Each of these properties relates to different strategies for enzyme immobilization. The main types of enzyme immobilization are adsorption, cross-linking, and entrapment/encapsulation.

Adsorption is the process of strong intermolecular forces resulting in the accumulation of protein on a solid surface. Adsorption is very dependent on the intermolecular interactions between the support surface and the enzyme. Therefore, properties like enzyme charge and polarity are crucially important to ensure high and reproducible coverage of enzyme on the support. This is typically a mild immobilization technique [15]. However, it is often problematic in that you typically have leaching of enzyme from the surface as a function of time.

Covalent binding and cross-linking are techniques that allow for the covalent binding of enzyme to surfaces or to other enzymes. These surfaces could be inner walls of a bioreactor or supports like glass or polymeric beads for a packed-bed reactor for industrial bioprocessing. Covalent binding or cross-linking of enzyme to enzyme to form aggregates is commonly referred to as CLEAs (cross-linked enzyme aggregates) [16] and they have been used in a number of applications. The cross-linking of protein to protein or protein to surface typically results in decreasing the enzyme activity of the protein, but it provides a great deal of stability to the protein. Covalent binding and cross-linking chemistry can sometimes be harsh chemistry. Therefore, it is important to consider the chemical microenvironment of the chemistry before use with proteins or enzymes, due to the fragility of the protein three-dimensional structure.

The final method of enzyme immobilization is entrapment or encapsulation within a polymeric matrix. Entrapment is the procedure of polymerizing a monomer or low molecular weight polymer around the protein to trap the protein on a surface. This is frequently done with sol-gels and is quite successful at immobilizing proteins on surfaces [17–20]. Traditional and electropolymerized polymers can also be employed for entrapping enzymes [21, 22]. Issues to consider with entrapment techniques are the chemical environment of the polymerization solution and whether it will denature the protein as well as the pore size and interconnectivity of the pores in the polymer to determine if substrate and product can diffuse in and out of the polymer, but also to ensure that the protein cannot diffuse out of the polymers. Polymer entrapment has been common, because many strategies do not result in a significant decrease in enzyme activity after immobilization. However, leaching is frequently a problem depending on enzyme loading and the entrapment technique tends to have less of an effect on enzyme stabilization at high temperatures and in organic solvents than cross-linking techniques. Encapsulation is similar to entrapment in that the protein is constrained within the polymeric matrix, but the polymer matrix has "pockets" or "pores" for immobilizing the protein. Micellar polymers are an example of polymers that can encapsulate an enzyme. The polymer micelles are swelled and the enzyme is allowed to intercalate into the micellar pockets/aggregates and the solvent is evaporated. This provides a polymer membrane that provides a chemical microenvironment similar to a cellular microenvironment that can provide temperature, pH, and organic solvent stability to the protein [23].

4 Applications

Immobilized and stabilized enzymes have a variety of applications. Stabilization is important for the commercialization of enzymes, because the enzymes must be stabilized for shipping and storage. Therefore, the small molecule additives are common for the commercial production and sale of enzymes of all quantities. However, stabilization is more difficult in industrial settings, so bioprocessing applications requiring enzymes to be immobilized on the inside of reactors or on supports in a packed bed reactor are typically more challenging than just stabilizing an enzyme for distribution and storage. Bioprocessing applications range from food applications (i.e., the use of glucose isomerase in the production high fructose of corn syrup) to pharmaceutical applications (i.e., enzymes to ensure chirality of drug products) to fuel applications (i.e., enzymatic production of butanol). These are frequently referred to as "cell-free biotechnologies" or "cell-free bioprocessing," because enzymes are subcellular components [24]. Other common applications of immobilized enzymes are biosensors and biofuel cells. Biosensors employ enzymes to selectively catalyze the reaction of an analyte [25]. Biofuel cells employ enzymes to catalyze the reduction of oxygen at the cathode of a fuel cell and/or the oxidation of fuel at the anode of a fuel cell [26]. The hostile pH and electric field environment in these applications results in a need for enzyme stabilization as well as immobilization on the electrode surface to improve efficiency. Overall, there is a wealth of industrial and laboratory applications for immobilized and stabilized enzymes. The following chapters provide protocols and examples for different strategies and techniques for immobilization and stabilization.

References

1. Hecky J, Muller KM (2005) Structural perturbation and compensation by directed evolution at physiological temperature leads to thermostabilization of beta-lactamase. Biochemistry 44:12640–12654
2. O'Fagain C (2003) Enzyme stabilization-recent experimental progress. Enzym Microb Technol 33:137–149
3. Liuu JH, Tsai FF, Liu JW, Cheng KJ, Cheng CL (2001) The catalytic domain of a Piromyces rhizinflata cellulase expressed in E. coli was stabilized by the linker peptide of the enzyme. Enzym Microb Technol 28:582–589
4. Matsura T, Miyai K, Trakulnaleamsai S, Yomo T, Shima Y, Miki S, Yamamoto K, Urabe I (1999) Evolutionary molecular engineering by random elongation mutagenesis. Nat Biotechnol 17:58–61
5. Jeng FY, Lin SC (2006) Characterization and application of PEGylated horseradish peroxidase for the synthesis of poly(2-naphthol). Process Biochem 41(7):1566–1573
6. Treetharnmathurot B, Ovartlarnporn C, Wungsintaweekul J, Duncan R, Wiwattanapatapee R (2008) Effect of PEG molecular weight and linking chemistry on the biological activity and thermal stability of PEGylated trypsin. Int J Pharm 357(1-2):252–259
7. Veronese FM (2001) Peptide and protein PEGylation: a review of problems and solutions. Biomaterials 22:405–417
8. Roberts MJ, Bentley MD, Harris JM (2002) Chemistry of peptide and protein PEGylation. Adv Drug Deliv Rev 54:459–476
9. Gomez L, Ramırez HL, Villalonga ML, Hernandez J, Villalonga R (2006) Immobilization of chitosan-modified invertase on alginate-coated chitin support via polyelectrolyte complex formation. Enzym Microb Technol 38:22–27
10. Martinek K, Klyachko NL, Kabanov AV, Khmel'nitskii YL, Levashov AV (1989) Micellar enzymology: its relation to membranology. Biochim Biophys Acta 981(2):161–172
11. Martinek K, Klyachko NL, Levashov AV, Berezin IV (1983) Micellar enzymology. Catalytic activity of peroxidase in a colloidal aqueous solution in an organic solvent. Dokl Akad Nauk USSR 263(2):491–493
12. Celej MS, D'Andrea MG, Campana PT, Fidelio GD, Bianconi ML (2004) Superactivity and conformational changes on chymotrypsin upon interfacial binding to cationic micelles. Biochem J 378:1059–1066
13. Cao L (2005) Immobilised enzymes: science or art? Curr Opin Chem Biol 9(2):217–226
14. Hanefeld U, Gardossi L, Magner E (2009) Understanding enzyme immobilisation. Chem Soc Rev 38:453–468
15. Cooney MJ, Svoboda V, Lau C, Martin GP, Minteer SD (2008) Enzyme catalysed biofuel cells. Energy Environ Sci 1:320–337
16. Kim MI, Kim J, Lee J, Jia H, Na HB, Youn JK, Kwak JH, Dohnalkova A, Grate JW, Wang P, Hyeon T, Park HG, Chang HM (2006) Crosslinked enzyme aggregates in hierarchically-ordered mesoporous silica: a simple and effective method for enzyme stabilization. Biotechnol Bioeng 96(2):210–218

17. Coche-Guerente L, Cosnier S, Labbe P (1997) Sol-gel derived composite materials for the construction of oxidase/peroxidase mediatorless biosensors. Chem Mater 9(6):1348–1352

18. Lim J, Malati P, Bonet F, Dunn B (2007) Nanostructured sol-gel electrodes for biofuel cells. J Electrochem Soc 154(2):A140–A145

19. Nguyen DT, Smit M, Dunn B, Zink JI (2002) Stabilization of creatine kinase encapsulated in silicate sol-gel materials and unusual temperature effects on its activity. Chem Mater 14:4300–4306

20. Hussain F, Birch DJS, Pickup JC (2005) Glucose sensing based on the intrinsic fluorescence of sol-gel immobilized yeast hexokinase. Anal Biochem 339:137–143

21. Yang R, Ruan Y, Deng J (1998) A H_2O_2 biosensor based on immobilization of horseradish peroxidase in electropolymerized methylene green film on GCE. J Appl Electrochem 28:1269–1275

22. Chiang C-J, Hsiau L-T, Lee W-C (2004) Immobilization of cell-associated enzymes by entrapment in polymethacrylamide beads. Biotechnol Tech 11(2):121–125

23. Moore CM, Akers NL, Hill AD, Johnson ZC, Minteer SD (2004) Improving the environment for immobilized dehydrogenase enzymes by modifying nafion with tetraalkylammonium bromides. Biomacromolecules 5(4):1241–1247

24. Bujara M et al (2010) Exploiting cell-free systems: implementation and debugging of a system of biotransformations. Biotechnol Bioeng 106:376–389

25. Aston WJ, Turner APF (1984) Biosensors and biofuel cells. Biotechnol Genet Eng Rev 1:89–120

26. Atanassov P, Apblett C, Banta S, Brozik S, Calabrese-Barton S, Cooney MJ, Liaw BY, Mukerjee S, Minteer SD (2007) Enzymatic biofuel cells. Interface 16(2):28–31

Chapter 2

Stabilization of Enzymes Through Encapsulation in Liposomes

Makoto Yoshimoto

Abstract

Phospholipid vesicle (liposome) offers an aqueous compartment surrounded by lipid bilayer membranes. Various enzyme molecules have been reported to be encapsulated in liposomes. The liposomal enzyme shows peculiar catalytic activity and selectivity to the substrate in the bulk liquid, which are predominantly derived from the substrate permeation resistance through the membrane. We reported that the quaternary structure of bovine liver catalase and alcohol dehydrogenase was stabilized in liposomes through their interaction with lipid membranes. The method and condition for preparing the enzyme-containing liposomes with well-defined size, lipid composition, and enzyme content are of particular importance, because these properties dominate the catalytic performance and stability of the liposomal enzymes.

Key words Liposomes, Phospholipid vesicles, Lipid bilayer membranes, Enzyme encapsulation, Membrane permeability, Enzyme structure, Enzyme reactivity, Bovine liver catalase

1 Introduction

The liposomal aqueous phase is isolated from the bulk liquid by the semipermeable lipid bilayer membranes, which means chemical reactions can be induced inside enzyme-containing liposomes by adding membrane-permeable substrate to the bulk liquid. In the liposomal system, the enzyme molecules are confined without chemical modification, which is advantageous to preserve the inherent enzyme affinity to the cofactor and substrate molecules. So far, various liposome-encapsulated enzymes have been prepared and characterized for developing diagnostic and biosensing materials, functional drugs, and biocompatible catalysts [1, 2]. The reactivity of liposomal enzymes was extensively examined mainly focusing on the membrane permeation of the substrate molecules as a rate-controlling step of the liposomal reaction [3, 4]. For example, sodium cholate is a useful modulator of the liposome membranes. Incorporation of sublytic concentrations of cholate in the membranes induced permeation of substrate and as a result the rate of liposomal enzyme

Shelley D. Minteer (ed.), *Enzyme Stabilization and Immobilization: Methods and Protocols*, Methods in Molecular Biology, vol. 1504, DOI 10.1007/978-1-4939-6499-4_2, © Springer Science+Business Media New York 2017

reaction increased [4]. An excess amount of cholate causes complete solubilization of liposome membrane, which is utilized for determining the total amount and inherent activity of the enzyme encapsulated in liposomes. On the other hand, the stability of enzyme activity in liposomes is relatively unknown. We recently reported that the thermostability of bovine liver catalase and yeast alcohol dehydrogenase considerably increased through encapsulation of each enzyme in liposomes composed of POPC (1-palmitoyl-2-oleoyl-*sn*-glycero-3-phosphocholine) [5–7]. Furthermore, the liposomal glucose oxidase system was shown to be applicable as a stable catalyst for the prolonged oxidation of glucose in the gas-liquid flow in a bubble column reactor [8–10]. In the liposomal system, the aggregate formation among the partially denatured enzyme molecules was indicated to be depressed through the interaction of the enzyme with lipid membranes [11, 12]. This chapter describes the preparation, reactivity and stability of the enzyme-containing liposomes with various sizes and enzyme contents using catalase as a model enzyme. The preparation and analytical methods described are basically applicable to liposomes containing other water-soluble enzymes. To prepare stable and reactive liposomal enzyme systems, the enzyme content in liposomes and the lipid composition need to be changed and optimized considering the characteristics of each enzyme employed and the permeability of its substrate through the liposome membranes.

2 Materials

2.1 Preparation of Catalase-Containing Liposomes

1. Phospholipid: 1-palmitoyl-2-oleoyl-*sn*-glycero-3-phosphocholine (POPC, >99%, $M_r = 760.1$, main phase transition temperature T_m of -2.5 ± 2.4 °C [13]) (Avanti Polar Lipids, Inc., Alabaster, AL).

2. Chloroform (>99%), diethylether (>99.5%).

3. Ethanol (>99.5%).

4. Dry ice (solid CO_2).

5. Rotary evaporator (REN-1, Iwaki Co. Ltd., Japan) with an aspirator (ASP-13, Iwaki Co. Ltd.).

6. Freeze-dryer (FRD-50 M, Asahi Techno Glass Corp., Funabashi, Japan) with a vacuum pump (GLD-051, ULVAC, Inc., Chigasaki, Japan).

7. Enzyme: bovine liver catalase (EC 1.11.1.6, ca. 10,000 U/mg, $M_r = 240,000$) (Wako Pure Chemical Industries, Ltd., Osaka, Japan).

8. Tris buffer: 50 mM Tris (2-amino-2-hydroxymethyl-1,3-propanediol)-HCl, pH 7.4, containing 0.1 M sodium chloride.

9. Small-volume extrusion device Liposofast™ and its stabilizer (Avestin Inc., Ottawa, Canada) [14].

10. Polycarbonate membranes for sizing liposomes (Avestin Inc., 19 mm in membrane diameter and 30, 50 or 100 nm in the nominal mean pore diameter).

11. Gel beads for gel permeation chromatography (GPC): sepharose 4B suspended in ethanol/water (GE Healthcare UK Ltd., Buckinghamshire, England).

12. Glass column with a stopcock for the GPC, 1.0 (id) × 35 cm, 20 mL in packed gel bed volume.

13. Enzyme kit for quantification of POPC (Phospholipid C-Test Wako, Wako Pure Chemical Industries, Ltd.).

14. UV/visible spectrophotometer (Ubest V-550DS, JASCO, Tokyo, Japan) equipped with a perche-type temperature controller (EHC-477S, JASCO).

2.2 Measurement of Enzyme Activity of Catalase-Containing Liposomes

1. 0.3 M sodium cholate (Wako Pure Chemical Industries, Ltd.) in Tris buffer.

2. Substrate of catalase: hydrogen peroxide (H_2O_2) solution (Wako Pure Chemical Industries, Ltd.).

3 Methods

3.1 Preparation of Catalase-Containing Liposomes

1. Weigh 50 mg of POPC powder (*see* **Note 1**).

2. Dissolve 50 mg of POPC in 4 mL of chloroform in a 100-mL round-bottom flask in a draft chamber.

3. Remove the solvent from the flask by using the rotary evaporator under reduced pressure in a draft chamber to form a lipid film on the inner wall of the flask.

4. Dissolve the lipid film in 4 mL of diethylether and remove the solvent as described above. Repeat this procedure once more.

5. Dry the lipid film formed in the flask by using the freeze-dryer connected to the vacuum pump for 2 h in the dark to remove the residual organic solvents molecules in the lipid layers. Keep the inner pressure of the flask less than 10 Pa throughout the drying process.

6. Dissolve the catalase in 2.0 mL of the Tris buffer in a glass test tube at the enzyme concentrations of 1.3–80 mg/mL (*see* **Note 2**).

7. Hydrate the dry lipid film with 2.0 mL of the enzyme-containing Tris buffer solution with gentle shaking to induce the formation of multilamellar vesicles (MLVs) (*see* **Note 3**).

8. Freeze the MLVs suspension with a refrigerant (dry ice/ethanol) in the Dewar flask for 5 min and then thaw it in a water bath thermostatted at 35 °C with gentle shaking. Repeat this freezing and thawing treatment seven times to transform a fraction of the small vesicles into the larger MLVs (*see* **Note 4**).

9. Extrude the MLVs suspension through a polycarbonate membrane with nominal mean pore diameter of 100, 50 or 30 nm 11 times using the extruding devise to obtain the catalase-containing unilamellar liposomes (*see* **Note 5**).

10. Pass the catalase-containing liposome suspension through the GPC column with the Tris buffer as an eluent collecting 1.0-mL fraction volumes in order to separate the catalase containing liposomes from the free (non-entrapped) enzyme molecules (*see* **Note 6**).

11. Measure the concentration of POPC in the liposome-containing fractions obtained by the chromatographic separation using the enzyme kit (*see* **Note 7**).

12. Store the enzyme-containing liposome suspension in a capped plastic tube in the dark at 4 °C until use (*see* **Note 8**).

3.2 Measurement of Enzyme Activity of Catalase-Containing Liposomes (See Tables 1 and 2)

1. Prepare 100 mM H_2O_2 in the Tris buffer solution considering the molar extinction coefficient of H_2O_2 at 240 nm (ε_{240}) is 39.4 M^{-1} cm^{-1} (*see* **Note 9**).

2. Add the catalase-containing liposome suspension to the Tris buffer solution containing H_2O_2 to give the intrinsic catalase and initial H_2O_2 concentrations of 0.1–0.2 μg/mL (about 1.0–2.0 U/mL) and 10 mM, respectively, and the total volume of 3.0 mL in a quartz cuvette in order to initiate the liposomal catalase-catalyzed decomposition of H_2O_2 (*see* **Note 10**).

3. Measure the time course of H_2O_2 decomposition at 25 ± 0.3 °C for 60 s based on the decrease in absorbance at 240 nm using the spectrophotometer. The decomposition rate of H_2O_2 is taken as the enzyme activity of liposomal catalase.

4. Measure the intrinsic enzyme activity of the liposomal catalase in the same way as above except that the reaction solution contains 40 mM sodium cholate for complete solubilization of liposome membranes (*see* **Note 11**).

5. Calculate the activity efficiency E of the liposomal catalase as its observed activity measured in the absence of cholate relative to the intrinsic one (*see* **Note 12**).

6. Calculate the catalase concentration in liposomes C_{in} and the number of biologically active catalase molecules per liposome N (*see* **Note 13**).

3.3 Stability of Liposomal Catalase Activity (See Figs. 1 and 2)

The stabilities of free and liposomal catalase at 55 °C are shown in Figs. 1 and 2, respectively [5]. It is clearly seen in Fig. 1 that the thermal stability of free catalase is dependent on its concentration. The higher the enzyme concentration employed, the higher its thermal stability at the free catalase concentration range of 0.25 µg/mL-5.0 mg/mL. This means that at the enzyme concentrations, dissociation of the enzyme into its subunits dominates the enzyme deactivation observed. On the other hand, the catalase at the highest concentration of 16 mg/mL shows lower stability than that at 5.0 µg/mL. For the 16 mg/mL of catalase, the formation of irreversible intermolecular aggregates is facilitated among the conformationally altered enzyme molecules. Figure 2 shows the stability of liposomal catalase systems with different liposomal enzyme concentrations C_{in} at 55 °C. The liposomal catalase with C_{in} of 4.9 mg/mL (CAL$_{100}$-II) shows the highest thermal stability. Quite importantly, the thermal stability of the liposomal catalase with C_{in} of 16 mg/mL (CAL$_{100}$-III) is much higher than that of free enzyme at the identical concentration (*see* Fig. 1). This is because the formation of enzyme aggregates is prevented in the liposomal aqueous phase through the interaction of the inner surface of the liposome membrane with the encapsulated enzyme molecules [5]. The liposomal system therefore can be a functional carrier that has the stabilization effect on the structure and activity of the liposome-encapsulated catalase molecules.

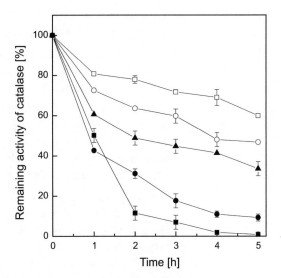

Fig. 1 Time courses of remaining activity of free catalase at enzyme concentrations of 16 mg/mL (*closed circles*), 5.0 mg/mL (*open squares*), 0.1 mg/mL (*open circles*), 5.0 µg/mL (*closed triangles*), and 0.25 µg/mL (*closed squares*) in 50 mM Tris–HCl/100 mM NaCl buffer (pH 7.4) at 55 °C (reproduced from Ref. [5] with permission from Elsevier Inc., Amsterdam)

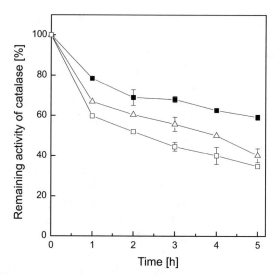

Fig. 2 Time courses of remaining activity of catalase in CAL$_{100}$-I (*open squares*), CAL$_{100}$-II (*closed squares*), and CAL$_{100}$-III (*open triangles*) at fixed POPC concentration of 2.8 mM in Tris buffer (pH 7.4) at 55 °C. The overall catalase concentrations were 5.9, 52, and 140 µg/mL for suspensions of CAL$_{100}$-I, CAL$_{100}$-II, and CAL$_{100}$-III, respectively. For detailed characteristics of CAL$_{100}$, *see* Table 1 (reproduced from Ref. [5] with permission from Elsevier Inc., Amsterdam)

4 Notes

1. Unsaturated lipid POPC is hygroscopic and thus should be treated under dry atmosphere.

2. Water should be sterilized and deionized using a water purification system (Elix 3UV, Millipore Corp., Billerica, MA). The minimum resistance to the water is 15 MΩ·cm.

3. Avoid vigorous mixing using a vortex mixer to minimize possible conformational change of the enzyme through its adsorption to the gas–liquid interface. The freezing and thawing treatments (*see* below) with occasional gentle shaking induce complete removal of the lipid film from the wall of the flask.

4. For preparing the refrigerant, the dry ice is crashed into small pieces with a hammer and mixed with ethanol in the Dewar flask. Check the effects of the repetitive freezing and thawing treatments on the enzyme activity. When saturated lipids with high T_m are employed instead of POPC, the thawing temperature should be carefully controlled to minimize the thermal deactivation or partial denaturation of the enzyme molecules.

5. The bubbles formed in the syringes of the extruder should be carefully removed before passing the MLV suspension through the membrane. The commercially available stabilizer for the

extruder is recommended to be used. For preparing the liposomes with mean diameter of about 30 or 50 nm, the MLVs are extruded through the pores with the diameter of 100 nm and then the membranes with smaller pores are employed for further reduction in the size of liposomes.

6. The GPC column is prepared by packing sepharose 4B gel beads into the glass column. Avoid the formation of bubbles in the gel bed. The ethanol originally contained in the gel suspension should be exhaustively removed by eluting the Tris buffer solution through the column to eliminate the effect of ethanol on the enzyme conformation. Using a transparent glass column is advantageous to visualize elution and separation behaviors of slightly colored enzymes including catalase and turbid liposomes. The sample volume applied to the GPC column is less than 1.0 mL. Elution of liposomes and subsequent free (non-entrapped) enzyme molecules is quantitatively confirmed by measuring the optical density at 400 nm and the enzyme activity, respectively of each fraction collected. For obtaining the enzyme-containing liposomes with narrow size distribution, the liposome-containing fractions should not be mixed together.

7. This assay is applicable to the determination of phospholipids containing choline group such as POPC. The POPC concentration in an enzyme-containing liposome suspension is selectively measured by quantifying the H_2O_2 produced from the lipid by a series of reactions catalyzed by phospholipase D, choline oxidase, and peroxidase. The effect of catalase activity derived from the catalase-containing liposomes is negligible in the quantification of the H_2O_2 produced because the enzyme concentration in the assay solution is low enough.

8. The deactivation of catalase molecules encapsulated in liposomes at the mean number of enzyme molecules per liposome of 5.2 is negligible at least for 22 d [5]. The storage stability of the liposomal catalase decreases with decreasing the liposomal enzyme concentration [9]. On the other hand, the stability of liposomal enzyme is practically unaffected by the liposome concentration.

9. Since decomposition of H_2O_2 slowly occurs during its storage, the 100 mM H_2O_2 solution should be freshly prepared each day.

10. The catalase-catalyzed decomposition of H_2O_2 yields oxygen and water. The stoichiometric equation is $H_2O_2 \rightarrow (1/2) O_2 + H_2O$.

11. Effects of cholate and cholate/lipid mixed micelles on the enzyme activity measurement should be checked. The activity of bovine liver catalase is practically unaffected by cholate up to

40 mM. The minimal cholate concentration required for complete solubilization of liposomes is generally dependent on the lipid concentration in the system and the lipid-water partitioning behavior of cholate as previously reported [3, 4]. Triton X-100 (*tert*-octylphenoxypolyethoxyethanol) can be alternatively used for solubilization of liposome membranes.

12. The observed activity of liposomal enzyme is smaller than the intrinsic one because of the permeation resistance of lipid membranes to the substrate (H_2O_2) molecules as shown in Table 1. The reactivity of other liposomal enzymes is controlled by modulating the preformed enzyme-containing liposomes with sublytic concentrations of detergents such as sodium cholate and Triton X-100 [4] and the channel-forming protein [10]. Mechanical stresses such as liquid shear and gas-liquid flow were also shown to be effective for increasing the permeability of liposome membranes to the dye molecules with low molecular mass [15–17]. To obtain a reliable *E* value in the presence of the membrane modulators, the leakage of enzyme molecules from the liposome interior to the bulk liquid should be minimized.

13. To determine the POPC liposome concentration, the number of lipid molecules per liposome is calculated on the basis of the assumptions that the bilayer membrane thickness *t* is 37 Å, the mean head group area *A* is 72 Å^2 [18], and liposomes are spherical with their diameter of *D*. The number of lipid molecules per liposome *n* can be calculated as $n = (4\pi/A)\{D^2/4 + (D/2 - t)^2\}$. The liposome concentration C_v is then determined as $C_v = C_T/n$, where C_T stands for the lipid concentration measured. The mean

Table 1
Characteristics of catalase-containing liposomes [5]

Liposomal catalase[a]	Concentration of catalase in hydration step C_0 [mg/mL]	Concentration of catalase in liposome C_{in} [mg/mL]	Mean number of catalase molecules in a liposome N [−]	Activity efficiency of liposomal catalase E [−]
CAL$_{30}$	1.3	66[b]	0.032 ± 0.008	0.59 ± 0.01
CAL$_{50}$	1.3	9.9[b]	0.15 ± 0.02	0.54 ± 0.06
CAL$_{100}$-I	1.3	0.96[b]	0.64 ± 0.06	0.57 ± 0.04
CAL$_{100}$-II	20	4.9 ± 0.9	5.2 ± 0.9	0.40 ± 0.04
CAL$_{100}$-III	80	16 ± 0.9	17 ± 0.9	0.34 ± 0.03

[a]CALs mean catalase-containing liposomes with the subscripts standing for the mean diameter of liposomes approximately equal to the nominal pore size used in the extrusion step
[b]C_{in} values were calculated assuming that each liposome contain one catalase molecule (reproduced from Ref. [5] with modification with permission from Elsevier Inc., Amsterdam)

Table 2
Characteristics of various enzyme-containing liposomes reported

Enzyme encapsulated in liposomes	Concentration of enzyme in hydration of step C_0 [mg/mL]	Concentration of enzyme in liposome C_{in} [mg/mL]	Ref.
α-chymotrypsin from *bovine pancreas*	20	7.1	[19]
Glucose oxidase from *Aspergillus niger*	1.3	0.78	[7]
Proteinase K from *Tritirachium album*	2.0	0.95	[4]
Yeast alcohol dehydrogenase	5.0	3.3	[6]

The C_{in} values are calculated assuming that the diameter D of the enzyme-containing liposomes is 115 [11] or 100 nm

number of active enzyme molecules per liposome N could be calculated as the concentration of active enzyme relative to that of liposomes C_v determined as described above. As shown in Table 1, the catalase concentration in liposomes C_{in} is definitely dependent on that in the Tris buffer C_0 employed in the lipid hydration (**step 7** in the Subheading 3.1). In the table the N values are less than unity for the catalase-containing liposomes prepared at C_0 of 1.3 mg/mL. This means that the liposomal enzyme suspensions prepared at the above condition are the mixture of catalase-containing liposomes and empty (enzyme-free) ones. For the liposomal catalase with $N > 1$, the C_{in} values are smaller than C_0. For other liposomal enzyme systems reported, the C_{in} values are consistently smaller than the C_0 ones, *see* Table 2 for details.

Acknowledgement

The author would like to thank Prof. Emeritus Katsumi Nakao (Yamaguchi Univ.) for his advice on biochemical engineering applications of enzyme-containing liposomes. This work was supported in part by the Japan Society of the Promotion of Science (no. 21760642).

References

1. Walde P, Ichikawa S (2001) Enzymes inside lipid vesicle: preparation, reactivity and applications. Biomol Eng 18:143–177

2. Walde P, Ichikawa S, Yoshimoto M (2009) The fabrication and applications of enzyme-containing vesicles (Chapter 7). In: Ariga K, Nalwa HS (eds) Bottom-up nanofabrication, 2nd edn. American Scientific Publishers, Stevenson Ranch, CA, pp 199–221

3. Treyer M, Walde P, Oberholzer T (2002) Permeability enhancement of lipid vesicles to nucleotides by use of sodium cholate: basic

studies and application to an enzyme-catalyzed reaction occurring inside the vesicles. Langmuir 18:1043–1050

4. Yoshimoto M, Wang S, Fukunaga K, Treyer M, Walde P, Kuboi R, Nakao K (2004) Enhancement of apparent substrate selectivity of proteinase K encapsulated in liposomes through a cholate-induced alterations of the bilayer permeability. Biotechnol Bioeng 85:222–233

5. Yoshimoto M, Sakamoto H, Yoshimoto N, Kuboi R, Nakao K (2007) Stabilization of quaternary structure and activity of bovine liver catalase through encapsulation in liposomes. Enzyme Microb Technol 41: 849–858

6. Yoshimoto M, Sato M, Yoshimoto N, Nakao K (2008) Liposomal encapsulation of yeast alcohol dehydrogenase with cofactor for stabilization of the enzyme structure and activity. Biotechnol Prog 24:576–582

7. Yoshimoto M, Miyazaki Y, Sato M, Fukunaga K, Kuboi R, Nakao K (2004) Mechanism for high stability of liposomal glucose oxidase to inhibitor hydrogen peroxide produced in prolonged glucose oxidation. Bioconjug Chem 15:1055–1061

8. Wang S, Yoshimoto M, Fukunaga K, Nakao K (2003) Optimal covalent immobilization of glucose oxidase-contaiing liposomes for highly stable biocatalyst in bioreactor. Biotechnol Bioeng 83:444–453

9. Yoshimoto M, Miyazaki Y, Kudo Y, Fukunaga K, Nakao K (2006) Glucose oxidation catalyzed by liposomal glucose oxidase in the presence of catalase-containing liposomes. Biotechnol Prog 22:704–709

10. Yoshimoto M, Wang S, Fukunaga K, Fournier D, Walde P, Kuboi R, Nakao K (2005) Novel immobilized liposomal glucose oxidase system using the channel protein OmpF and catalase. Biotechnol Bioeng 90:231–238

11. Kuboi R, Yoshimoto M, Walde P, Luisi PL (1997) Refolding of carbonic anhydrase assisted by 1-palmitoyl-2-oleoyl-sn-glycero-3-phosphocholine liposomes. Biotechnol Prog 13:828–836

12. Yoshimoto M, Kozono R, Tsubomura N (2015) Liposomes as chaperone mimics with controllable affinity toward heat-denatured formate dehydrogenase from Candida boidinii. Langmuir 31:762–770

13. Walde P (2004) Preparation of vesicles (liposomes). In: Nalwa HS (ed) Encyclopedia of nanoscience and nanotechnology, 2nd edn. American Scientific Publishers, Stevenson Ranch, CA, pp 43–79

14. MacDonald RC, MacDonald RI, Menco BPM, Takeshita K, Subbarao NK, Hu L-R (1991) Small-volume extrusion apparatus for preparation of large, unilamellar vesicles. Biochim Biophys Acta 1061:297–303

15. Yoshimoto M, Monden M, Jiang Z, Nakao K (2007) Permeabilization of phospholipid bilayer membranes induced by gas-liquid flow in an airlift bubble column. Biotechnol Prog 23:1321–1326

16. Natsume T, Yoshimoto M (2014) Mechanosensitive liposomes as artificial chaperones for shear driven acceleration of enzyme-catalyzed reaction. ACS Appl Mater Interfaces 6:3671–3679

17. Natsume T, Yoshimoto M (2013) A method to estimate the average shear rate in a bubble column using liposomes. Ind Eng Chem Res 52:18498–18502

18. Dorovska-Taran V, Wick R, Walde P (1996) A ^{1}H nuclear magnetic resonance method for investigating the phospholipase D-catalyzed hydrolysis of phosphatidylcholine in liposomes. Anal Biochem 240:37–47

19. Yoshimoto M, Walde P, Umakoshi H, Kuboi R (1999) Conformationally changed cytochrome c-induced fusion of enzyme- and substrate-containing liposomes. Biotechnol Prog 15:689–696

Chapter 3

Micellar Enzymology for Thermal, pH, and Solvent Stability

Shelley D. Minteer

Abstract

This chapter describes methods for enzyme stabilization using micellar solutions. Micellar solutions have been shown to increase the thermal stability, as well as the pH and solvent tolerance of enzymes. This field is traditionally referred to as micellar enzymology. This chapter details the use of ionic and nonionic micelles for the stabilization of polyphenol oxidase, lipase, and catalase, although this method could be used with any enzymatic system or enzyme cascade system.

Key words Micellar enzymology, Micelles, Superactivity, Enzyme stabilization

1 Introduction

Micellar enzymology is a technique for improving the thermal, pH, or solvent stability of enzymes using micelles [1]. These micelles provide a more protein–friendly, chemical microenvironment for protecting the enzyme. It has long been said that micelles provide a chemical microenvironment for the protein/enzyme that is more like the cellular chemical microenvironment than any traditionally employed buffer solution [2]. Incorporating enzymes into micelles can be done in three different ways: (1) the injection method, (2) the phase transfer method, and (3) direct solubilization. In the most popular method, a small volume of enzyme dissolved in an aqueous solution is injected into a surfactant organic solvent solution [3]. This injection method forms a reverse micelle where the non-polar group orientates toward the organic solvent and the polar groups orientate to the inside of the vesicle and traps an aqueous pool inside the micelle that contains the enzyme. The phase transfer method is similar in that the surfactant is dissolved in an organic solvent and enzyme is dissolved in an aqueous solution and the two solutions are mixed vigorously. The final method, direct solubilizing, is used with the enzyme is not soluble in water. In this case, the lyophilized enzyme is added directly into a water–organic solvent surfactant

Shelley D. Minteer (ed.), *Enzyme Stabilization and Immobilization: Methods and Protocols*, Methods in Molecular Biology, vol. 1504, DOI 10.1007/978-1-4939-6499-4_3, © Springer Science+Business Media New York 2017

solution and mixed until the solution becomes a transparent micellar solution and the enzyme powder is dissolved.

The size of the micelle can be tailored by altering the relative amounts of water and surfactant. If you increase the concentration of water, you will increase the volume or size of the reverse micelles. If you increase the concentration of surfactant, you will decrease the size and volume of the reverse micelles. This can be important for optimizing the micelle for each individual enzyme.

Micelles can be formed from a variety of surfactants including Aerosol-OT, sodium bis(2-ethylhexyl)sulfosuccinate (AOT), cetyl-trimethylammonium bromide (CTAB), Triton X-100, and fatty acids (i.e., oleic acid, linolenic acid, linoleic acid). Each of the micelles alters the chemical microenvironment as well as altering the size and distribution of the reverse micelles. Unfortunately, since little is known about the relationship between protein structure and ideal chemical microenvironment, optimization of micellar size, charge, and hydrophobicity is required for each protein employed. The following protocols are examples of protein–micelle combinations, but many options can be employed. The key to optimization of the protein–micelle combination is to have an activity assay that is appropriate for the enzyme and is not affected by the chemical and physical properties of the micelle.

2 Materials

2.1 AOT Micelles for Superactivity of Polyphenol Oxidase

1. Mushroom polyphenol oxidase (Sigma).
2. AOT (sodium salt, Sigma).
3. Cyclohexane.
4. Phosphate buffer: 25 mM sodium phosphate buffer (pH 6.5).

2.2 Stabilization of a Membrane Protein in Lipid Micelles

1. Mitochondrial malate dehydrogenase (Sigma) (*see* **Note 1**).
2. Sodium hydroxide solution: 2.5 mM.
3. Biosonic Ultrasonic.
4. Stearic acid (sodium salt).
5. Oleic acid (sodium salt).
6. Alcohol, reagent grade.

2.3 Nonionic/Cationic Surfactant Micelles for Stabilization of Lipase

1. Lipase from *Chromobacterium viscosum* (Sigma).
2. CTAB (Sigma).
3. Brij 30 and Brij 92 (Sigma).
4. 4-nitrophenyloctanoate (Sigma) (*see* **Note 2**).
5. Phosphate buffer: 20 mM phosphate buffer (pH 6.0).

6. Enzyme solution: 0.34 mg lipase/mL distilled and deionized water.

7. Isooctane.

8. *n*-hexanol, reagent grade.

2.4 Extend the Temperature Stability at 50 °C of Catalase in Nonionic Reverse Micelles

1. Bovine-liver catalase (Sigma).

2. Brij 30 (Sigma).

3. *n*-heptane.

4. Phosphate buffer: 10 mM phosphate buffer (pH 7.0).

3 Methods

3.1 AOT Micelles for Superactivity of Polyphenol Oxidase [4]

This method is based on work described by Rojo et al. in Ref. [4]. As described above, there are three main methods for incorporating enzymes into micelles, the method used here is the phase transfer method. The micelle surfactant AOT is dissolved in the organic solvent cyclohexane followed by the introduction of buffer to form the micelles. After the micelles are formed, a buffered solution of enzyme is added to incorporate the polyphenol oxidase. This results in the kinetics for phenol increasing from 1.19 ± 0.04 μmol/(min*mg) for aqueous polyphenol oxidase to 3.77 ± 0.10 μmol/(min*mg) for the AOT reverse micelles. In the micellar enzymology field, this is referred to as superactivity.

1. A 50 mM solution of AOT in cyclohexane was prepared.

2. 75 μl of pH 6.5 buffer was added to 5 mL of the AOT solution and mixed rigorously (*see* **Note 3**).

3. 9 μg of polyphenol oxidase was dissolved in 15 μl of pH 6.5 buffer.

4. Then, 15 μl of the enzyme solution is mixed vigorously with the 5 mL of reverse micelles formed in **step 2**. The final Wo for the micelles should be around 20 (*see* **Note 4**).

3.2 Stabilization of a Membrane Protein in Lipid Micelles [5]

Membrane proteins are inherently unstable in aqueous solutions. Mitochondrial malate dehydrogenase is a prime example of this. This protein is an oxidoreductase enzyme of the Kreb's cycle that is responsible for oxidizing malate to oxaloacetate or vice versa. Mitochondrial malate dehydrogenase loses all activity within 7 h if dissolved in pH 7.4 Tris–acetate buffer. This instability is unacceptable for almost any biotechnology application. The protein is a small hydrophobic protein that has been shown to require hydrophobic protein–protein interactions in the mitochondrial structure to maintain its stability. Lipid micelles have been shown to provide a bio-inspired method for stabilization of membrane proteins, like mitochondrial malate dehydrogenase. This protocol describes the stabilization of malate dehydrogenase in stearic acid and oleic acid micelles. Both micelles have resulted in superactivity of the malate dehydrogenase compared to malate dehydrogenase in buffer. Oleic acid micelles result in a 40 % increase in malate dehydrogenase activity, whereas the stearic acid micelles only result in a 25 % increase in activity.

1. Stearic acid micelles were formed by sonicating a solution of 2.5 mM NaOH with an amount of fatty acid corresponding to a final concentration of 25 μM. The sonication should be done for 2–3 min at 90 kilo-cycles on a Biosonik Ultrasonic. The resulting solution should be homogeneous and opalescent.

2. A solution of oleic acid in 95 % alcohol was formed. A dispersion is formed by injecting the oleic acid solution into 2.5 mM NaOH with an amount of oleic acid corresponding to a final concentration of 12 μM. This should result in a homogeneous and clear solution.

3.3 Nonionic/Cationic Surfactant Micelles for Stabilization of Lipase [6]

Lipase is an industrially relevant esterase enzyme that hydrolyzes ester bonds. It is frequently used as a test enzyme, because of biological and industrial relevance. This protocol describes the stabilization of lipase in nonionic/cationic micelles. Brij is a class of nonionic surfactants that are poly(oxyethylene) alkyl ethers. Some are used for simples micelles and some for reverse micelles [7]. Brij 30 and 92 is commonly used for reverse micelles in nonpolar, organic solvents. The cationic surfactant used in this study is CTAB, which is an ammonium salt that is well known to form surfactant micelles.

1. CTAB and either Brij 30 or Brij 92 were added to 2 mL of isooctane, so the final concentration is 25 mM CTAB and 25 mM of the nonionic surfactant (Brij 30 or Brij 92).

2. n-hexanol is added to the solution in a proportion to ensure that that ratio of [n-hexanol] to [surfactant] is between 4.8 and 6.4.

3. This solution must be shaken vigorously and thoroughly mixed before adding pH 6.0 phosphate buffer. Phosphate buffer is added to reach a desired ratio of [water]–[surfactant] of 4:24.

4. This solution is vortexed until homogeneous.

5. 4.5 μl of enzyme solution is added to 1.5 mL of the reverse micelle solution above (*see* **Note 5**).

3.4 Extend the Temperature Stability at 50 °C of Catalase in Nonionic Reverse Micelles [7]

Catalase is an oxidoreductase enzyme that catalyzing the reaction of hydrogen peroxide to water and oxygen. It is a common enzyme and has been used here to show the ability to gain increased temperature tolerance with the use of Brij 30 micelles. As discussed above, Brij 30 micelles are nonionic micelles that have been used frequently for studying micellar affects and enzyme stabilization.

1. A stock solution of 0.1 M Brij 30 in n-heptane is prepared.

2. An aqueous solution of 10^{-8} M catalase enzyme in pH 7.0 10 mM phosphate buffer is injected into the stock Brij 30 solution in proportions so the ratio of $[H_2O]/[\text{surfactant}]$ is 4.5.

3. The mixture should be shaken vigorously until a completely transparent solution results.

4 Notes

1. Mitochondrial malate dehydrogenase can also be isolated from whole pig heart according to a method in Ref. [8].

2. Nitrophenyloctanoate can also be synthesized as described in Ref. [6].

3. Simply pipetting the buffer into the AOT solution will not form micelles. It is important for vigorous mixing to occur. Vortexers or other mechanical mixers are suggested.

4. Simply pipetting the buffer into the reverse micelle solution is not sufficient for incorporation of the protein. It is important for vigorous mixing to occur. Vortexers or other mechanical mixers are suggested.

5. This solution should become clear within 1 min of gentle shaking or there is a problem with the reverse micelle preparation.

References

1. Martinek K, Klyachko NL, Kabanov AV, Khmel'nitskii YL, Levashov AV (1989) Micellar enzymology: its relation to membranology. Biochim Biophys Acta 981(2):161–172

2. Shipovskov S, Levashov AV (2004) Entrapping of tyrosinase in a system of reverse micelles. Biocatal Biotransformation 22(1):57–60

3. Chang GG, Huang TM, Hung HC (2000) Reverse micelles as life-mimicking systems. Proc Natl Sci Counc Repub China B 24(3):89–100

4. Rojo M, Gomez M, Isorna P, Estrada P (2001) Micellar catalysis of polyphenyl oxidase in AOT/cyclohexane. J Mol Catal B Enzym 11:857–865

5. Callahan JW, Kosicki GW (1967) The effect of lipid micelles on mitochondrial malate dehydrogenase. Can J Biochem 45:839–851

6. Shome A, Roy S, Das PK (2007) Nonionic surfactants: a key to enhance the enzyme activity at cationic reverse micellar interface. Langmuir 23:4130–4136

7. Gebicka L, Jurgas-Grudzinska M (2004) Activity and stability of catalase in nonionic micellar and reverse miceller systems. Z Naturfosch 59c:887–891

8. Kitto GB, Wasserman PM, Michjeda J, Kaplan NO (1966) Multiple forms of mitochondrial malate dehydrogenases. Biochem Biophys Res Commun 22:75–81

Chapter 4

Lipase Activation and Stabilization in Room-Temperature Ionic Liquids

Joel L. Kaar

Abstract

Widespread interest in the use of room-temperature ionic liquids (RTILs) as solvents in anhydrous biocatalytic reactions has largely been met with underwhelming results. Enzymes are frequently inactivated in RTILs as a result of the influence of solvent on the enzyme's microenvironment, be it through interacting with the enzyme or enzyme-bound water molecules. The purpose of this chapter is to present a rational approach to mediate RTIL–enzyme interactions, which is essential if we are to realize the advantages of RTILs over conventional solvents for biocatalysis in full. The underlying premise for this approach is the stabilization of enzyme structure via multipoint covalent immobilization within a polyurethane foam matrix. Additionally, the approach entails the use of salt hydrates to control the level of hydration of the immobilized enzyme, which is critical to the activation of enzymes in nonaqueous media. Although lipase is used as a model enzyme, this approach may be effective in activating and stabilizing virtually any enzyme in RTILs.

Key words Nonaqueous biocatalysis, Ionic liquids, Enzyme stabilization, Immobilization, Salt hydrates, Water activity, Lipase, Green chemistry

1 Introduction

As solvents for anhydrous biocatalysis, room-temperature ionic liquids (RTILs) have generated overwhelming excitement due to their combination of highly desirable properties [1–10]. The signature properties of RTILs include thermal stability, nonexistent vapor pressure, and remarkable physiochemical flexibility, the latter of which is due to the interchangeability of the cation and anion. The intrinsic flexibility of RTILs is particularly attractive for chemical processing since RTILs can be rationally designed to dissolve a broad range of potential substrates, hence the designation of RTILs as *designer solvents*. However, to fully exploit the perceived benefits of RTILs for biocatalysis, strategies for mediating the interaction between solvent and enzyme are required.

To date, the use of enzymes in neat RTILs has met with only limited success. The RTILs which support enzyme activity are, in

Shelley D. Minteer (ed.), *Enzyme Stabilization and Immobilization: Methods and Protocols*, Methods in Molecular Biology, vol. 1504, DOI 10.1007/978-1-4939-6499-4_4, © Springer Science+Business Media New York 2017

almost all cases, those with a dialkyl imidazolium cation with hexa-fluorophosphate, tetrafluoroborate, trifluoromethylsulfonate, or bis(trifluoromethylsulfonyl)imide anions. The modest results can be largely attributed to the propensity of RTILs to associate with enzymes, which is not surprising given their ionic composition. Structure–activity data from our earlier work suggests the interaction of enzymes and RTILs is anion dependent [11]. Anions that are strongly nucleophilic such as nitrate, acetate, and trifluoroacetate may coordinate with positively charged sites in an enzyme, resulting in deactivating conformation changes [12–19]. Likewise, RTILs may interact with enzymes through hydrogen bonding networks, which can similarly evoke changes in enzyme structure. Micaêlo and Soares [20] have previously showed via molecular simulation that the nitrate anion of RTILs forms hydrogen bonds with amide groups in the backbone of cutinase.

Solvent effects on enzyme hydration, which influences activity, stability, and specificity in nonaqueous environments [21, 22], are equally important in RTILs. This is because RTILs are highly hygroscopic and, as such, may absorb significant amounts of water. The adsorption of water, should it exceed the holding capacity of (i.e., saturate) the RTIL, may facilitate denaturation of the enzyme at the solvent–water interface and thus reduce enzyme activity. Moreover, it is plausible that RTILs may compete for catalytically essential water from an enzyme in the same manner organic solvents of similarly high polarity and hydrophilicity do [20, 23]. We have previously reported on the impact of thermodynamic water activity (a_w), a measure of the amount of enzyme-associated water, on enzyme activity in the RTIL 1-butyl-3-methylimidazolium hexafluorophosphate ([bmim][PF$_6$]) [24].

Incorporation of enzymes into polymers is a conventional method of enzyme stabilization that may prevent deactivating conformation changes in RTILs. Enzyme-containing polymers are prepared by reacting a prepolymer with an enzyme, resulting in multipoint covalent immobilization of the enzyme within the polymer network. In this way, the enzyme plays the role of a monomer in the polymerization reaction. The formation of linkages between enzyme and polymer restricts the mobility of the enzyme, thereby suppressing its unfolding due to environmental pressures. Polyurethanes in particular are attractive supports due to their rapid and simple preparation and tunable properties that, when conjugated to enzymes, yield stable bioplastics [25–31]. Of course, stabilization of enzyme structure alone will not suffice to markedly improve biocatalytic efficiency if there is too much or too little water associated with the enzyme or if inactivation takes place via a mechanism that does not involve unfolding, such as oxidation. We have also recently shown that RTIL-induced inactivation may be prevented by altering the presence of ionizable groups on the surface of enzymes [32–35]. The alteration of surface electrostatics may improve enzyme tolerance to RTILs via weakening ionic and hydrophobic interactions with the cation and anion at critical sites.

In this chapter, a twofold method for improving the activity and stability of lipase, a classic model enzyme that catalyzes esterification and transesterification reactions in anhydrous media, in RTILs is described (Fig. 1a). First, lipase is immobilized within a polyurethane foam matrix as a means of stabilizing enzyme structure in RTILs. Synthesis of lipase-containing foams is based on the reaction between amines, and to a lesser extent hydroxyl groups, on the enzyme's surface with a toluene diisocyanate prepolymer (Fig. 1b). Second, salt hydrates are employed to modulate a_w in the RTIL. By assaying the activity of immobilized lipase at fixed a_w in

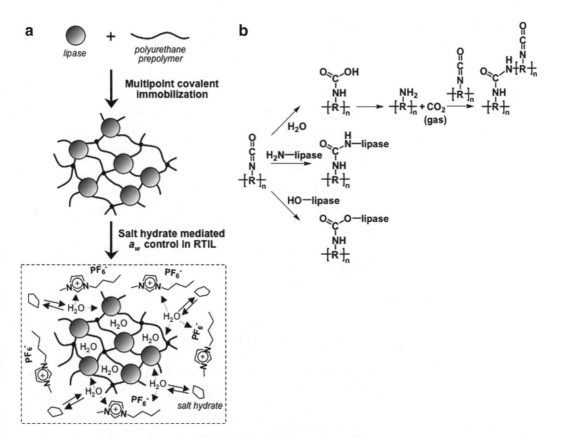

Fig. 1 Strategy for lipase activation and stabilization in [bmim][PF₆]. (**a**) Lipase is initially immobilized covalently within a polyurethane foam matrix. The a_w of the ionic liquid milieu is then modulated by salt hydrates, which, via interconversion of the salt between higher and lower hydration states, create a dynamic water equilibrium. Water is distributed to the individual reaction components, including enzyme and solvent, such that the thermodynamic availability of water to each is equivalent. (**b**) Reaction scheme of lipase-immobilized polyurethane foam synthesis (adapted from Ref. [39]). Isocyanate groups in the prepolymer react with water to generate carbamic acid, which upon elimination of carbon dioxide, is transformed into an amine. The free amine, along with primary amines and hydroxyl functionalities in the enzyme, further react with the isocyanate groups to produce a polymer network. The production of carbon dioxide during synthesis causes the polymer to expand into a highly porous foam

RTILs, the optimal hydration level of the enzyme can be ascertained. The lipase-catalyzed transesterification of methyl methacrylate and 2-ethylhexanol is a convenient model reaction system for this purpose. Additionally, stability studies that measure lipase activity in water after pre-incubation in RTILs are useful in determining if the effects of RTILs on enzymes are reversible or permanent. Although not discussed in this chapter, this method may be combined with rational modification of enzyme charge to further enhance enzyme stability in RTILs.

2 Materials

2.1 Reagents

1. Lipase (*Candida antarctica*) type B (Novozym CALB L; Novozymes, Denmark). The enzyme does not require further purification. Store at −20 °C.

2. Immobilization buffer: 50 mM *bis–tris* propane, 5 mM calcium chloride, pH 7.5 containing 1 % (w/w) Pluronic L62, a nonionic surfactant (BASF, Parsippany, NJ, USA).

3. Hypol 3000, a water-miscible polyurethane prepolymer of polyethylene glycol with toluene diisocyanate reactive groups (Dow Chemical, Midland, MI, USA) (*see* **Note 1**).

4. [bmim][PF$_6$] (SACHEM, Austin, TX, USA) (*see* **Note 2**).

5. Transesterification substrates: methyl methacrylate and 2-ethylhexanol (*see* **Note 3**).

6. 2-ethylhexyl methacrylate standard for following the lipase-catalyzed transesterification of methyl methacrylate and 2-ethylhexanol by gas chromatography.

7. Hexane.

8. Salt hydrates: sodium phosphate dibasic dihydrate (Na$_2$HPO$_4 \cdot 2$H$_2$O); sodium acetate trihydrate (NaAC $\cdot 3$H$_2$O); copper (II) sulfate pentahydrate (CuSO$_4 \cdot 5$H$_2$O); sodium pyrophosphate decahydrate (Na$_4$P$_2$O$_7 \cdot 10$H$_2$O); sodium phosphate dibasic heptahydrate (Na$_2$HPO$_4 \cdot 7$H$_2$O); sodium phosphate dibasic dodecahydrate (Na$_2$HPO$_4 \cdot 12$H$_2$O) (Table 1). Salt hydrate pairs are referred to in shorthand notation as the chemical formula of the salt followed by the number of water molecules associated with the higher and lower hydrate (i.e., Na$_2$HPO$_4$ 2/0 refers to the mixture of Na$_2$HPO$_4 \cdot 2$H$_2$O and Na$_2$HPO$_4 \cdot 0$H$_2$O).

9. Lipase stability assay buffer: 200 mM Tris buffer, pH 7.7.

10. Olive oil lipase substrate emulsion (Sigma, St. Louis, MO, USA): 4.5 mM triolein, 1 M NaCl, 13 % (w/v) Triton X-100. Store at 4 °C.

11. 95 % (v/v) ethanol.

12. 0.9 % (w/v) thymolphthalein solution.

13. 0.05 N sodium hydroxide.

Table 1
Water activities of salt hydrate pairs in [bmim][PF$_6$] at 25 °C as measured using a humidity sensor. The values were compared to those of the salt hydrate pairs compiled from the literature from both experiment and simulation. (Reproduced from Ref. [24]) with permission from John Wiley & Sons, Ltd.)

	a_w		
Salt Hydrate Pairs	Ref. [43]	Ref. [44]	[bmim][PF$_6$]
NaI 2/0	0.12	0.07	0.17
Na$_2$HPO$_4$ 2/0	0.16	-	0.17
NaAc 3/0	0.28	0.34	0.28
CuSO$_4$ 5/3	0.32	0.42	0.42
Na$_4$P$_2$O$_7$ 10/0	0.49	0.47	0.47
Na$_2$HPO$_4$ 7/2	0.61	0.61	0.63
Na$_2$HPO$_4$ 12/7	0.80	0.79	0.78

2.2 Equipment

1. Handheld drill and metal oar-shaped mixing head (height 3.2 cm × diameter 1.3 cm).

2. Sandpaper (grit no. 180).

3. Sonicating water bath.

4. Orbital shaking incubator.

5. 1-μL Hamilton syringe.

6. Gas chromatograph equipped with an Alltech (Deerfield, IL, USA) Econo-Cap 1000 capillary column (length 30 m × inner diameter 0.53 mm × film thickness 1.0 μm) and a flame ionization detector.

7. 25-mL glass burette with 0.1 mL graduations.

3   Methods

3.1 Polyurethane-Lipase Polymer Synthesis

1. Initially, add 1 g of lipase to 2.5 mL of immobilization buffer in a 50-mL conical plastic tube.

2. Add 2.5 g of Hypol 3000 to the lipase solution, resulting in a final concentration of 1 g-prepolymer/mL-buffer.

3. Immediately after adding the polyurethane prepolymer, vigorously mix the biphasic reaction solution using the handheld drill for 30 s, or until there is no remaining residual prepolymer, at 2500 rpm. The resulting polymer is a homogenous foam, which will expand several times in volume. The timescale for the polymerization reaction to reach completion is on the order of 10 min (*see* **Note 4**).

4. Allow the lipase-containing polymer to dry overnight in the reaction tube, uncovered, at ambient temperature and pressure. After drying, the foam should be stored at 4 °C in order to preserve enzymatic activity (*see* **Note 5**).

5. In preparation or use in transesterifcation reactions, grind the enzyme-polymer foam into small (submillimeter size) particles using the handheld drill with cylindrical bit wrapped with the sandpaper. The sandpaper can be attached to the bit using double-sided tape or any adhesive material. Low rpm drill speed should be employed in attempt to achieve a constant particle size (*see* **Note 6**).

3.2 Lipase-Catalyzed Transesterification of Methyl Methacrylate and 2-Ethylhexanol in [bmim][PF₆] at Fixed a_w

1. Prepare substrate solutions (2 mL) of methyl methacrylate and 2-ethylhexanol in [bmim][PF$_6$] at a concentration of 200 mM. Considering the density of [bmim][PF$_6$] (1.38 g/mL), the solutions are prepared by adding 42.8 μL of methyl methacrylate to 2.70 g of [bmim][PF$_6$] and 62.5 μL of 2-ethylhexanol to 2.67 g of [bmim][PF$_6$]. The solutions should be thoroughly mixed, which can be achieved using a magnetic stirrer.

2. Pre-equilibrate the reaction components with the salt hydrate to obtain the desired a_w. Typically, 20–80 mg free or immobilized lipase and 0.4 g salt hydrate are added to 1 mL of the 2-ethylhexanol solution in a 5-mL Kimble glass vial. In a separate vial, 0.8 g of the same salt hydrate is added to 2 mL of methyl methacrylate solution. The vials are incubated for 1 h at 30 °C and 300 rpm in the orbital shaker to achieve the equilibrium a_w (*see* **Note 7**).

3. After pre-equilibration, initiate the reaction by adding 1 mL of the methyl methacrylate to 1 mL of the solution of 2-ethylhexanol containing enzyme and salt hydrate. Briefly (15–30 s) sonicate the reaction suspension to disperse the enzyme particles after which place the vial containing the reaction in the orbital shaker set at 30 °C and 300 rpm.

4. Sample the reaction by periodically withdrawing 50 μL (69 mg) aliquots from the reaction mixture over 1–5 h. Add 100 μL of hexane to the aliquot and, using a pipette, repeatedly aspirate (~10 times) to extract the transesterification product.

5. Monitor the initial rate of 2-ethylhexyl methacrylate formation by gas chromatography (GC) analysis of the extraction solution. For the purpose of this method, it is assumed the capillary column has been properly installed in the GC and conditioned and that the hydrogen and air flows are adjusted to the optimum range and ratio. Helium is used as the carrier gas with a pressure setting of 6.5 psi. Typically, 0.5–1.0 μL of the extraction solution is injected into the GC using a 1:4 split ratio to

dilute the sample. The injector and detector should be set at 300 °C. The oven program consists of an initial temperature of 100 °C, which is maintained for 2 min. The temperature is then ramped to 160 °C at a rate of 25 °C/min and held for 3 min. Under these conditions, the retention time for 2-ethyl-hexyl methacrylate is 5.1 min. The concentration of 2-ethyl-hexyl methacrylate can be determined from the calibration of peak area versus 2-ethylhexyl methacrylate concentration. The calibration should be made in the presence of the substrates to closely match the reaction sampling conditions (*see* **Note 8**).

6. Enzyme activity is typically expressed as the rate of 2-ethyl-hexyl methacrylate formation per unit mass of free or immobi-lized lipase in the reaction.

3.3 Lipase Stability in [bmim][PF₆]

1. Add 10 mg free or immobilized lipase to a 5-mL Kimble glass vial and subsequently cover with 1 mL (1.38 g) [bmim][PF$_6$]. Prepare one vial of the enzyme suspension for each time point to be taken.

2. Incubate the enzyme suspension at a specified temperature (i.e., 30 or 50 °C) while agitating at 300 rpm in the orbital shaker.

3. At each time point, extract the solvent from the vial using a pipette with care taken to minimize the loss of enzyme.

4. Assay the hydrolytic activity of the enzyme by measuring the rate with which it catalyzes the conversion of olive oil triglycer-ides to fatty acids in buffer. Typically, 2.5 mL of deionized water, 1 mL of lipase stability assay buffer, and 3 mL of the olive oil substrate were added to the vial containing the enzyme. The reaction solution is incubated at 50 °C for 2 h while shaken at 300 rpm. After incubation, transfer the solution to a 50-mL flask to which 3 mL of 95 % (v/v) ethanol is then added. To the flask add approximately 40 μL of the thymolphthalein indicator solution and subsequently swirl to mix. Titrate the reaction solu-tion by adding 0.05 N sodium hydroxide dropwise from the burette, stopping to swirl after every few drops, until the solu-tion turns a faint blue color, marking the titration endpoint (*see* **Note 9**). Lipase activity is defined as the volume of titrant added per unit mass of enzyme in the vial at each time point from which that at the time zero incubation point is subtracted. The activity at each time point may be normalized to the time zero point, thereby yielding relative enzyme stability (*see* **Note 10**).

4 Notes

1. Care should be taken to prevent water from being introduced to Hypol 3000 since water will initiate its polymerization. Accordingly, it is recommended that Hypol 3000 is stored at room temperature in a desiccator.

2. RTILs should be free of residual halide ions, which can significantly reduce enzyme activity. The presence of impurities including halides in RTILs is easily detectable since such impurities alter the color of RTILs and, in some cases, impart an odor (RTILs, when pure, are colorless and odorless). The purity of dialkylimidazolium and pyrrolidinium RTILs can be readily confirmed by UV–Vis spectroscopy [36]. The characteristic spectra of these RTILs do not include any significant absorption peaks above 350 nm. If required, halide impurities or others can be removed from RTILs using activated carbon or silica gel [36–38]. Additionally, because RTILs are hygroscopic and thus absorb water over time, [bmim][PF$_6$] should also be stored at room temperature in a desiccator. Water absorption can significantly alter the chemical and physical properties of RTILs including density, viscosity, and dielectric constant. If required, RTILs may be dried under vacuum to remove residual water.

3. All solvents, including substrates and hexane, should be of the highest purity possible to prevent introducing impurities that may inhibit lipase activity and potentially complicate GC analysis by giving rise to additional peaks.

4. Longer mixing times may be required when scaling up the immobilization reaction. The required mixing time will also differ when using prepolymers with differing hydrophobicity [39]. General, the required mixing time increases with increasing hydrophobicity of the prepolymer. Prepolymers of different hydrophobicity may be beneficial when preparing polyurethane foam-immobilized enzymes for different reaction environments (i.e., different RTILs). The hydrophobicity of the prepolymer dictates that of the resulting foam. Changes in polymer hydrophobicity impact the partitioning of reaction substrates and products from the bulk solution to the foam, thus effecting enzyme activity. Hence, it is feasible to enhance the activity of polyurethane-immobilized biocatalysts in nonaqueous media by tuning the hydrophobicity of the polymer matrix.

5. The extent of lipase immobilization can be determined by assaying enzyme leaching from the polyurethane foam. To do so, rinse blocks of the lipase-immobilized foam in a large volume (100–500 mL) of immobilization buffer, with stirring, for several hours (~4–6 h minimum) and subsequently assay protein concentration of the rinsate. Protein concentration can be quantified using the Bradford [40] or bicinchoninic acid [41] methods or spectrophotometrically by measuring protein absorption at 280 nm [42].

6. Alternatively, the foam can be cut into small pieces using a sharp blade or scissors.

7. Only the higher salt hydrate is added to the substrate solutions since the lower hydrate is formed by the release of water. The amount of salt hydrate added is variable depending on the solubility of the salt in the RTIL. Residual amounts of the solid salts are required for a_w control. The time required to reach the equilibrium a_w can vary depending on the salt hydrate used [24]. For Na_2HPO_4 7/2, NaAc 3/0, and Na_2HPO_4 2/0, a 1 h incubation time is sufficient to achieve the equilibrium a_w in [bmim][PF_6]. On the contrary, longer equilibration times are required for the salt hydrate pairs Na_2HPO_4 12/7 (4 h) and $Na_4P_2O_7$ 10/0 (>20 h).

8. To ensure the settings of the GC have not changed with time, which may alter the retention times and peak size of the reaction products, the GC should be calibrated daily by a single point calibration; otherwise an internal standard should be used.

9. It is best to do the titration against a white backdrop so as to make to the color of the titration solution easily identifiable. As the titration endpoint is neared, the solution may appear blue upon addition of the indicator dye only to become translucent after swirling. The titration should not be stopped until a blue tint remains after swirling.

10. Our results found that immobilized forms of lipase retained nearly full activity upon incubation in [bmim][PF_6] over 24 h at 30 °C [11]. Immobilized lipase was considerably less stable in RTILs containing [bmim] and 1-methyl-1-(-2-methoxyethyl) pyrrolidinium ([mmep]) cations with the nitrate anion at the same temperature. Interestingly, incubation in select RTILs, including [mmep][CH_3SO_3], [bmim][CH_3CO_2], and, [mmep] [CH_3CO_2], resulted in a significant increase in lipase (Novozym 435) activity when returned to water. The increased activity of the immobilized lipase may be explained by swelling of the immobilization support in these RTILs, causing previously inaccessible enzyme to be solvent exposed and thus an increase in active site concentration. While seemingly less likely, it is also plausible that the RTIL exposed enzyme has a modified tertiary structure with increased activity in water.

Acknowledgments

This work was funded by SACHEM Inc. and by a research grant from the Environmental Protection Agency (R-82813101-0) to Alan J. Russell, to whom I am grateful for support, both financial and scientific. I am thankful to Jason A. Berberich for technical advice and helpful discussion on all aspects of this work. I wish to also thank Richard R. Koepsel for critical reading of the manuscript.

References

1. Brennecke JF, Maginn EJ (2001) Ionic liquids: innovative fluids for chemical processing. AIChE J 47:2384–2389

2. Yang Z, Pan W (2005) Ionic liquids: green solvents for nonaqueous biocatalysis. Enzyme Microb Technol 37:19–28

3. van Rantwijk F, Sheldon RA (2007) Biocatalysis in ionic liquids. Chem Rev 107:2757–2785

4. Roosen C, Muller P, Greiner L (2008) Ionic liquids in biotechnology: applications and perspectives for biotransformations. Appl Microbiol Biotechnol 81:607–614

5. Dominguez de Maria P, Maugeri Z (2011) Ionic liquids in biotransformations: from proof-of-concept to emerging deep-eutectic-solvents. Curr Opin Chem Biol 15:220–225

6. Mai NL, Ahn K, Bae SW, Shin DW, Morya VK, Koo YM (2014) Ionic liquids as novel solvents for the synthesis of sugar fatty acid ester. Biotechnol J 9:1565–1572

7. Sun S, Qin F, Bi Y, Chen J, Yang G, Liu W (2013) Enhanced transesterification of ethyl ferulate with glycerol for preparing glyceryl diferulate using a lipase in ionic liquids as reaction medium. Biotechnol Lett 35:1449–1454

8. Carter JL, Bekhouche M, Noiriel A, Blum LJ, Doumeche B (2014) Directed evolution of a formate dehydrogenase for increased tolerance to ionic liquids reveals a new site for increasing the stability. Chembiochem 15:2710–2718

9. Patel DD, Lee JM (2012) Applications of ionic liquids. Chem Rec 12:329–355

10. Bi YH, Duan ZQ, Li XQ, Wang ZY, Zhao XR (2015) Introducing biobased ionic liquids as the nonaqueous media for enzymatic synthesis of phosphatidylserine. J Agric Food Chem 63:1558–1561

11. Kaar JL, Jesionowski AM, Berberich JA, Moulton R, Russell AJ (2003) Impact of ionic liquid physical properties on lipase activity and stability. J Am Chem Soc 125:4125–4131

12. Geng F, Zheng LQ, Yu L, Li GZ, Tung CH (2010) Interaction of bovine serum albumin and long-chain imidazolium ionic liquid measured by fluorescence spectra and surface tension. Process Biochem 45:306–311

13. Lai JQ, Li Z, Lu YH, Yang Z (2011) Specific ion effects of ionic liquids on enzyme activity and stability. Green Chem 13:1860–1868

14. Shu Y, Liu ML, Chen S, Chen XW, Wang JH (2011) New Insight into molecular interactions of imidazolium ionic liquids with bovine serum albumin. J Phys Chem B 115:12306–12314

15. Singh T, Bharmoria P, Morikawa M, Kimizuka N, Kumar A (2012) Ionic liquids induced structural changes of bovine serum albumin in aqueous media: a detailed physicochemical and spectroscopic study. J Phys Chem B 116:11924–11935

16. Singh T, Boral S, Bohidar HB, Kumar A (2010) Interaction of gelatin with room temperature ionic liquids: a detailed physicochemical study. J Phys Chem B 114:8441–8448

17. Klahn M, Lim GS, Seduraman A, Wu P (2011) On the different roles of anions and cations in the solvation of enzymes in ionic liquids. Phys Chem Chem Phys 13:1649–1662

18. Burney PR, Pfaendtner J (2013) Structural and dynamic features of *Candida rugosa* lipase 1 in water, octane, toluene, and ionic liquids BMIM-PF6 and BMIM-NO3. J Phys Chem B 117:2662–2670

19. Jaeger VW, Pfaendtner J (2013) Structure, dynamics, and activity of xylanase solvated in binary mixtures of ionic liquid and water. ACS Chem Biol 8:1179–1186

20. Micaelo NM, Soares CM (2008) Protein structure and dynamics in ionic liquids. Insights from molecular dynamics simulation studies. J Phys Chem B 112:2566–2572

21. Halling PJ (1994) Thermodynamic predictions for biocatalysis in nonconventional media: theory, tests, and recommendations for experimental design and analysis. Enzyme Microb Technol 16:178–206

22. Halling PJ (2004) What can we learn by studying enzymes in non-aqueous media? Philos Trans R Soc Lond B Biol Sci 359:1287–1296, discussion 1296–1287, 1323–1288

23. Nakashima K, Maruyama T, Kamiya N, Goto M (2006) Homogeneous enzymatic reactions in ionic liquids with poly(ethylene glycol)-modified subtilisin. Org Biomol Chem 4:3462–3467

24. Berberich JA, Kaar JL, Russell AJ (2003) Use of salt hydrate pairs to control water activity for enzyme catalysis in ionic liquids. Biotechnol Prog 19:1029–1032

25. Lejeune KE, Mesiano AJ, Bower SB, Grimsley JK, Wild JR, Russell AJ (1997) Dramatically stabilized phosphotriesterase-polymers for nerve agent degradation. Biotechnol Bioeng 54:105–114

26. Gordon RK, Feaster SR, Russell AJ, LeJeune KE, Maxwell DM, Lenz DE, Ross M, Doctor BP (1999) Organophosphate skin decontamination using immobilized enzymes. Chem Biol Interact 119–120:463–470

27. LeJeune KE, Swers JS, Hetro AD, Donahey GP, Russell AJ (1999) Increasing the tolerance of organophosphorus hydrolase to bleach. Biotechnol Bioeng 64:250–254

28. Gill I, Ballesteros A (2000) Bioencapsulation within synthetic polymers (Part 2): non-sol-gel protein-polymer biocomposites. Trends Biotechnol 18:469–479

29. Drevon GF, Hartleib J, Scharff E, Ruterjans H, Russell AJ (2001) Thermoinactivation of diisopropylfluorophosphatase-containing polyurethane polymers. Biomacromolecules 2:664–671

30. Drevon GF, Danielmeier K, Federspiel W, Stolz DB, Wicks DA, Yu PC, Russell AJ (2002) High-activity enzyme-polyurethane coatings. Biotechnol Bioeng 79:785–794

31. Vasudevan PT, Lopez-Cortes N, Caswell H, Reyes-Duarte D, Plou FJ, Ballesteros A, Como K, Thomson T (2004) A novel hydrophilic support, CoFoam, for enzyme immobilization. Biotechnol Lett 26:473–477

32. Nordwald EM, Armstrong GS, Kaar JL (2014) NMR-guided rational engineering of an ionic liquid-tolerant lipase. ACS Catal 4:4057–4064

33. Nordwald EM, Brunecky R, Himmel ME, Beckham GT, Kaar JL (2014) Charge engineering of cellulases improves ionic liquid tolerance and reduces lignin inhibition. Biotechnol Bioeng 111:1541–1549

34. Nordwald EM, Kaar JL (2013) Stabilization of enzymes in ionic liquids via modification of enzyme charge. Biotechnol Bioeng 110:2352–2360

35. Nordwald EM, Kaar JL (2013) Mediating electrostatic binding of 1-butyl-3-methylimidazolium chloride to enzyme surfaces improves conformational stability. J Phys Chem B 117:8977–8986

36. Burrell AK, Del Sesto RE, Baker SN, McCleskey TM, Baker GA (2007) The large scale synthesis of pure imidazolium and pyrrolidinium ionic liquids. Green Chem 9:449–454

37. Park S, Kazlauskas RJ (2001) Improved preparation and use of room-temperature ionic liquids in lipase-catalyzed enantio- and regioselective acylations. J Org Chem 66:8395–8401

38. Lee SH, Ha SH, Lee SB, Koo YM (2006) Adverse effect of chloride impurities on lipase-catalyzed transesterifications in ionic liquids. Biotechnol Lett 28:1335–1339

39. Lejeune KE, Russell AJ (1996) Covalent binding of a nerve agent hydrolyzing enzyme within polyurethane foams. Biotechnol Bioeng 51:450–457

40. Bradford MM (1976) A rapid and sensitive method for the quantitation of microgram quantities of protein utilizing the principle of protein-dye binding. Anal Biochem 72:248–254

41. Smith PK, Krohn RI, Hermanson GT, Mallia AK, Gartner FH, Provenzano MD, Fujimoto EK, Goeke NM, Olson BJ, Klenk DC (1985) Measurement of protein using bicinchoninic acid. Anal Biochem 150:76–85

42. Pace CN, Vajdos F, Fee L, Grimsley G, Gray T (1995) How to measure and predict the molar absorption coefficient of a protein. Protein Sci 4:2411–2423

43. Halling PJ (1992) Salt hydrates for water activity control with biocatalysts in organic media. Biotechnol Tech 6:271–276

44. Zacharis E, Omar IC, Partridge J, Robb DA, Halling PJ (1997) Selection of salt hydrate pairs for use in water control in enzyme catalysis in organic solvents. Biotechnol Bioeng 55:367–374

Chapter 5

Nanoporous Gold for Enzyme Immobilization

Keith J. Stine, Kenise Jefferson, and Olga V. Shulga

Abstract

Nanoporous gold (NPG) is a material of emerging interest for immobilization of biomolecules, especially enzymes. The material provides a high surface area form of gold that is suitable for physisorption or for covalent modification by self-assembled monolayers. The material can be used as a high surface area electrode and with immobilized enzymes can be used for amperometric detection schemes. NPG can be prepared in a variety of formats from alloys containing between 20 and 50 % atomic composition of gold and less noble element(s) by dealloying procedures. Materials resembling NPG can be prepared by hydrothermal and electrodeposition methods. Related high surface area gold structures have been prepared using templating approaches. Covalent enzyme immobilization can be achieved by first forming a self-assembled monolayer on NPG bearing a terminal reactive functional group followed by conjugation to the enzyme through amide linkages to lysine residues. Enzymes can also be entrapped by physisorption or immobilized by electrostatic interactions.

Key words Nanoporous gold, Porous gold, Enzyme immobilization, Self-assembled monolayer, Bioconjugation

1 Introduction

Nanoporous gold (NPG) has recently joined the variety of gold nanostructures upon which the immobilization of biomolecules, especially enzymes, is being pursued for applications in sensors and assays [1–5], supported synthesis [6, 7], catalysis [5, 8], fuel cells [9, 10], and biofuel cells [11, 12]. The structure of NPG consists of interconnected ligaments and pores of typical average width 20–200 nm [13–15]. A micrograph of NPG is shown in Fig. 1. NPG combines the attractive features of high surface area, suitability for modification using self-assembled monolayers (SAMs), use as a high surface area electrode, potential to be prepared as a supported film or as a free-standing structure and possessing a tunable pore size over a range of several nm up to about a micron. A recently published monograph has surveyed research on NPG [16].

NPG can be prepared by a number of methods, with the most common being the treatment of alloys containing 20-50 atomic

Shelley D. Minteer (ed.), *Enzyme Stabilization and Immobilization: Methods and Protocols*, Methods in Molecular Biology, vol. 1504, DOI 10.1007/978-1-4939-6499-4_5, © Springer Science+Business Media New York 2017

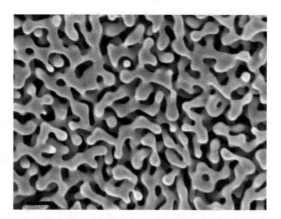

Fig. 1 Electron micrograph of nanoporous gold obtained using a low-voltage field emission scanning electron microscope (JEOL JSM 6320F) at a magnification of 100,000× and a voltage of 5 kV. The scale bar in the lower left corner represents 200 nm. The sample was obtained by dealloying in nitric acid as reported in ref. [40]

percent gold in a strong acid such as nitric acid to achieve a process known as dealloying in which all elements other than gold are oxidized and removed by dissolution [13–16]. The process of dealloying can also be achieved electrochemically by application of a potential positive enough to oxidize the less noble element(s) present in the alloy [17–20]. The average pore size of the NPG produced depends upon the applied potential [21, 22] and also upon the dealloying temperature [23]. Dealloying at lower temperature produces lower average pore diameters, and so does dealloying at higher applied potentials. At percentages of gold above 50%, the porous structure does not form and only surface pitting occurs, and at too low a percentage of gold the material will fall apart upon dealloying [24]. Although the most commonly chosen alloys from which to form NPG are gold + silver alloys due to the nearly identical lattice parameters of gold and silver and their ideal miscibility, NPG has also been obtained by treatment of other gold containing alloys such as Au–Zn [25], Au–Sn [26, 27], Au–Cu [28–31], Au–Al [32–36], a ternary Au–Pd–Ag alloy [37] and a multicomponent jewelry alloy [23]. Dealloying of ternary alloys of Au–Pd–Ag [38] produced a nanoporous Au–Pd alloy, and dealloying of Ag–Au–Pt alloy with 6% Pt produced NPG with pores below 5 nm in size [39]. Surface areas for NPG reported by BET (Brunauer–Emmett–Teller) nitrogen gas adsorption isotherm analysis are typically 4–10 m^2/g [40, 41]. Pore size distributions have been reported based on the Barrett–Joyner–Halenda analysis [42].

The mechanism of dealloying has been described as a spinodal decomposition process occurring at the alloy-solution interface [43, 44]. During the earliest stages of dealloying, the interface consists of gold islands and exposed alloy. As the ions of the less noble element(s) diffuse away into solution, the gold islands become undercut as ridges and ligaments begin to form. This process proceeds into the alloy until the final structure of randomly interconnected Au ligaments is obtained. The three dimensional structure of NPG has been confirmed using transmission electron tomography [13]. The average pore size of NPG can be adjusted by annealing at elevated temperature with longer annealing time resulting in a larger increase in the average pore size [15, 42, 45]. Exposure to acid for longer periods also results in a gradually increasing average pore size [14]. While such annealing decreases the surface area, it is of interest for studying the effects of pore size on molecular diffusion and accommodation of biomolecules of different sizes inside NPG.

The number of studies of enzyme immobilization on NPG is steadily growing. Studies of enzyme immobilization on microporous gold electrodes prepared by deposition of gold onto colloidal particle templates [46–48] and on gold nanowires [49, 50] or nanotubes [51] prepared using template methods are a related group of studies. Some studies that have been stated as being on porous gold are actually on highly rough forms of gold or "gold black" [52]. A process of multiple potential scans applied to a gold electrode in a $ZnCl_2$ + benzyl alcohol solution was reported to produce a material resembling NPG [53]. In another approach, reduction of an Au salt by formaldehyde following by treatment at high temperature resulting in a material resembling NPG [54]. Materials prepared by other hydrothermal [55] or direct electrodeposition procedures without dealloying often resemble interconnected porous networks of nanoparticles [56, 57]. Electrodeposition from 0.1 M $HAuCl_4$ in 0–3 M NH_4Cl yielded a foam-like morphology of Au with a multimodal pore size [57]. In this chapter, recent approaches to enzyme immobilization on NPG and some closely similar Au nanomaterials are described. Studies using these materials are in their early stages and much remains to be learned about the optimal usage of NPG for enzyme immobilization before optimal protocols can be confidently prescribed. For example, the distribution of protein immobilized inside macroscopic pieces of NPG was examined by atomic force microscopy and was found to depend on whether immobilization occurred under static or flow-through conditions [58].

Studies of the immobilization of enzymes on NPG obtained by dealloying have recently appeared for enzymes laccase [11, 12, 55, 56, 59–61], bilirubin oxidase [62, 63], glucose oxidase [64–66], acetylcholinesterase [40, 67], lipase [68, 69], lignin peroxidase [70], xylanase [71], horseradish peroxidase [69], catalase [69], cutinase [72], alkaline phosphatase conjugated to IgG monoclonal antibody [4, 73] and also to the lectin Concanavalin A [74]. Redox proteins whose immobilization has been studied include photosystem I [32], hemoglobin [54], and cytochrome c [36, 62]. Methods of immobilization used have included physisorption, electrostatic immobilization, and conjugation to self-assembled monolayers bearing a terminal functional group such as a carboxylic acid. Knowledge of the dimensions of the enzyme and the location of the lysine residues and the active site is of value towards understanding the possible orientation(s) of the enzyme at the NPG surface. Lysine residues are especially important as they may provide a fairly strong immobilization simply through attractive gold—amine interactions or can be used for covalent conjugation to carboxylic acid-terminated SAMs. NPG pore size obtained in many preparations of the material falls in a range of 20–200 nm which should be large enough to readily accommodate entry and diffusion of many enzymes for immobilization. Furthermore, the subsequent diffusion of small molecule substrates in and products out of the NPG is possible. The interior of NPG presents surfaces of primarily positive curvature along the ligaments but also regions of negative curvature near branches in the ligament structure presenting a different environment for enzyme immobilization. Many details concerning biomolecule immobilization within NPG remain to be studied. Below, efforts to immobilize enzymes on NPG are summarized according to some of the enzymes that have been of greatest interest.

1.1 Laccase

Laccase, a multicopper oxidase which acts on many substrates and concomitantly reduces oxygen to water, is important in developing cathodes for biofuel cells [75]. Laccase was immobilized onto NPG prepared from alloy foils of either 25 μm or 100 nm thickness that were dealloyed in concentrated nitric acid. Dealloying of 100 nm foils referred to commonly as "gold leaf," produces a structure with a small number of spanning ligaments [14]. In this study, the thin NPG was used to construct an electrode, while the thicker NPG was used for studies of enzyme loading and stability [11]. NPG of pore size 10–20 nm was found to result from 1 h acid treatment, while a 40–50 nm average pore size resulted from 17 h acid treatment. Annealing at 200 °C for 1 h increased the average pore size to 90–100 nm. The amount of laccase adsorbed was determined by applying the Bradford assay [79] to the supernatant solution and washings after immobilization to determine the decrease in enzyme concentration that could be attributable to its loading onto the NPG. Application of such a solution depletion approach clearly requires exposure of the volume of enzyme solution to an NPG sample of sufficient surface area such that the solution concentration is diminished to a readily measured extent. NPG samples allow for kinetics of the immobilized enzyme to be studied using standard methods and determination of the Michaelis–Menten parameters K_m and V_{max}, and also k_{cat} provided that the amount of immobilized enzyme has been determined. It was found that the amount of laccase loaded was lower for the sample with 10–20 nm pore size due to restricted entry of 7 nm diameter hydrated laccase while the loading onto the other two samples was similar. Activity was measured spectrophotometrically by following the decrease in absorbance at 470 nm using the substrate 2,6-dimethoxyphenol. Measurement of the mass loading of enzyme into NPG via the Bradford assay was combined with measurements of the rate of substrate conversion to yield enzyme activity for the immobilized enzyme. The sample of largest pore size was the most subject to loss of laccase by leaching. The thermal stability of the immobilized enzyme was much improved over that of the enzyme in solution. The electrocatalytic reduction of oxygen was clearly observed for the thin NPG fixed onto a glassy carbon electrode surface and the electrode response was unchanged after 1 month of storage. In a following study [12], the same lab compared physisorption with electrostatic immobilization and covalent immobilization. The enzyme loading achieved by electrostatic immobilization was half that of physisorption or covalent immobilization which were found to be similar. The similarity of enzyme loading for physisorption to covalent coupling to a lipoic acid self-assembled monolayer was accounted for by noting that laccase has 8 lysine residues that could form strong associations with a gold surface resulting in physisorption having a significant covalent nature of its own. The specific activity of the enzyme was found to be the same for all three methods. NPG prepared in the form of free-standing plates is a somewhat brittle material that can be easily crushed into fragments or pulverized into micron–sized particles. Comparison of the kinetic parameters for a full plate with crushed and pulverized plates revealed that K_m decreased for smaller NPG fragment size. The value of K_m for immobilized enzymes inside porous materials is often interpreted as an effective K_m influenced by restricted diffusion of substrate molecules within the pores.

Laccase was physisorbed onto NPG by drop casting a solution of the enzyme followed by drying under vacuum and then entrapped by applying a hydrophilic epoxy coating over the electrode surface [60]. The NPG was prepared by dealloying an Ag + Au alloy that had been co-sputtered onto an Au layer on a Ti-adhesion layer primed glass. An onset potential for catalytic reduction of oxygen near 650 mV (vs. Ag|AgCl) was observed, which agreed with the potential expected for the T1 copper site of the enzyme. The catalytic activity of the enzyme was about 40% higher at 37 °C than at 25 °C. If the enzyme was drop cast from a solution that also contained a polymer with side-chains bearing osmium complexes, the activity was threefold greater, attributed to an improved efficiency of electron transfer.

Immobilization of laccase by physisorption onto 4-aminothiophenol modified NPG was found to increase enzyme thermal stability [61]. At 50 °C and pH 5, the free enzyme lost 70% of its activity in 2 h, the enzyme on NPG alone lost 20%, and the enzyme on the SAM-modified NPG lost 10% of its activity. On NPG, immobilized laccase showed one reversible broad pair of oxidation and reduction peaks centered near 0.13 V (vs. SHE), while on the modified NPG a pair of broad reversible peaks centered near +0.26 and +0.55 V were observed. The presence of the 4-aminothiophenol was concluded as promoting electron transfer from the T1 copper site in laccase.

A material referred to as "highly porous gold" (hPG) was produced by two-step electrodeposition from $HAuCl_4$ solution onto gold yielding a morphology like that of an open-celled foam and possessing a very broad pore size distribution ranging from 10 nm to 30 µm [56]. A novel method of slow potential cycling (1 mV/s between 0.42 and 0.60 V, vs. SCE) was found to yield greater immobilization of glucose oxidase directly onto the material than simple physisorption. While the K_m value for the glucose oxidase on hPG was lower than that in solution (6.3 mM vs. 27 mM), activity was seen to decrease in 24 h by 35% attributed to coarsening of the Au nanostructure. Laccase was immobilized onto hPG that was first modified to present amine groups by adsorption of p-phenyldiazonium salt whose reduction was stated as yielding an Au–phenyl bond, followed by reduction of the terminal nitro groups to amines. The combination of the two enzyme-modified hPG electrodes was used to produce a biofuel cell with a power density of 6 µW/cm² at 0.2 V (vs. SCE) in 27.8 mM glucose in aerated pH 7 phosphate-buffered saline.

A macroporous gold film, prepared by drop casting a solution of gold "supraspheres," each of ~500 nm diameter prepared in solution by reduction of $HAuCl_4$ in the presence of bovine serum albumin, onto a fluorine doped indium tin oxide electrode followed by calcining at 450 °C, was used for immobilization of laccase [55]. The Au supraspheres were composed of 20 nm Au nanoparticles held together by protein. The resulting pore sizes were in the 1–3 µm range. Immobilization of laccase to create the biocathode was achieved by electrochemical reduction of p-nitrophenyl diazonium salt onto the surface to form an Au–phenyl bond, followed by reduction of the nitro groups to amines. The surface was then immersed in 6-mercapto-1-hexanol to fill in uncovered Au spots. Laccase was immobilized using EDC/NHS coupling, with the laccase first exposed to $NaIO_4$ to oxidize sugars on laccase glycans.

NPG prepared by co-sputtering of a 300 nm thick $Ag_{67}Au_{33}$ layer onto a 100 nm thick Au layer adhered to glass by a 10 nm thick Ti adhesion layer followed by acid dealloying was used for the immobilization of both cytochrome c and bilirubin oxidase from *Myrothecium verrucaria* (*Mv*BOD) [63]. Bilirubin oxidase is another multicopper oxidase important in biofuel cell development [75]. Cytochrome c was used as a test enzyme for comparing direct electron transfer on NPG and flat Au, and the *Mv*BOD was noted as useful for creating a biocathode. The cytochrome c

enzyme exhibited minor deviations from ideal behavior when covalently immobilized by amide linkage through lysines onto mixed SAMs with terminal carboxylic acid groups. The surface coverage of enzyme was enhanced over that on a flat Au electrode of the same geometric area (0.28 cm radius) by factors of 9–11; however, a significant fraction of the NPG surface was found to be not accessible for enzyme modification. The immobilization of *Mv*BOD on NPG by drop casting it from a solution containing an osmium-based redox polymer, a cross-linker, and a mediator was successful and yielded stable and substantial currents (500 $\mu A/cm^2$ in unstirred solution) for oxygen reduction to water.

Bilirubin oxidase was also immobilized on NPG by physisorption achieved by placing a drop of the enzyme in solution on the NPG electrode supported on glass [63] and then drying under vacuum. After enzyme physisorption, a hydrophilic epoxy layer was applied to entrap the enzyme. Direct electron transfer was observed in oxygen saturated 0.1 M citrate-phosphate buffer at pH 7. It was concluded that the oxygen reduction reaction was diffusion limited. The onset potential for O_2 reduction near 500 mV (vs. Ag|AgCl (sat'd KCl)) was consistent with the expected potential of 460 mV for the T1 active site of the enzyme.

The behavior of both laccase and glucose oxidase immobilized by adsorption onto NPG electrodes of 40 nm average pore diameter first surface modified with 4-aminothiophenol was investigated in air-saturated buffer of pH 5 to create a miniature biofuel cell [59]. The amount of enzyme adsorbed was determined by applying the Bradford assay to the supernatant. Electrochemical behavior was compared with and without a 4-aminothiophenol SAM on the NPG using cyclic voltammetry. Laccase showed broad reversible oxidation and reduction peaks on the SAM-modified NPG but with some peak splitting attributed to different environments for two of the copper centers; in contrast, no peak splitting was seen for the enzyme physisorbed on NPG. For glucose oxidase, oxidation and reduction peaks were seen only on the SAM-modified NPG. Use of the laccase and glucose oxidase electrodes together to make a biofuel cell gave a power density of 52 $\mu W/cm^2$ at 0.21 V (vs. SHE) in 100 mM glucose at pH 5. The enzymes were more stable on NPG than on flat gold due to the confinement effect provided by NPG.

1.2 Glucose Oxidase   Given the long and ongoing interest in electrochemical glucose sensors, immobilization of glucose oxidase on NPG has also begun to be pursued. The covalent immobilization of glucose oxidase on NPG was studied as a function of pore size [64]. NPG was prepared from alloy leaf that was $Au_{35}Ag_{65}$ in composition. The dealloyed leaf was supported on glassy carbon electrodes. The NPG was surface modified in tetrahydrofuran solution with the reactive linker dithiobis(succinimidyl undecanoate) for 10 h followed by linking to glucose oxidase (1 mg/ml in PBS for 5 h). The amperometric response of the electrodes to H_2O_2 at -0.2 V was found to be greatest for the smaller pore size of 18 nm vs. pore sizes of 30, 40, and 50 nm. For glucose sensing, the best response was for the 30 nm pore size. Glucose response was linear over the range 3–8 mM and the detection limit was 10 μM, with negligible interference from uric acid (0.1 mM) or ascorbic acid (20 μM). The enzyme electrode retained 90 % of its activity after 4 weeks of storage.

NPG prepared by dealloying 100 nm thick 12 carat Au leaf was attached to a glassy carbon electrode and then modified with a SAM of cysteamine, cross-linked with glutaraldehyde and then used to covalently immobilize glucose oxidase [65]. The performance as a glucose sensor was examined using two different mediators to shuttle electrons between the enzyme cofactor FAD in the active site and the electrode surface. It was found that p-benzoquinone, for which both the oxidized and reduced forms are neutral and inner sphere electron transfer occurs on the Au surface, gave almost threefold greater sensitivity to glucose than use of ferrocene carboxylic acid as a mediator, which undergoes outer sphere electron transfer. The lower sensitivity when ferrocence carboxylic acid (1.35 vs. 3.53 μA/cm^2/mM) was used as a mediator was attributed to electrostatic repulsion of the positively charged ferrocenium form. Determination of diffusion coefficients for the mediators led to the conclusion that only the outer regions of the NPG were involved in these fast redox reactions.

Given that it is known that NPG can itself promote the electrocatalytic oxidation of glucose initiated by hydroxide adsorption on gold [76], the possible synergistic oxidation of glucose by NPG and by glucose oxidase immobilized on NPG was investigated [66]. Glucose oxidase was physisorbed onto NPG of average pore diameter 35 nm, in accordance with a theory [77] that enzyme stabilization would be maximal in a cage that is 2–6× the protein size. The NPG electrode, prepared by dealloying 12 carat Au leaf and then attaching to glassy carbon electrode, was immersed in glucose oxidase (2000 U/ml, where U = unit of activity defined as oxidizing 1 mM β-D-glucose to D-glucolactone and H_2O_2 per minute at 35 °C and pH 5.1) in PBS (50 mM, pH 6.8) for 72 h. The glucose oxidase loaded NPG showed an oxidation peak near 0.4 V and a reduction peak near 0.3 V (vs. SCE) in deaerated PBS in the absence of glucose. Addition of glucose resulted in large oxidation peaks at both of these potentials. Such peaks were not seen using the NPG electrode alone. At 0.3 V, the peak current density was linear with glucose concentration over the range 50 μM–10 mM, and a high sensitivity of 12.1 mA/cm^2/mM was observed along with a detection limit of 1.02 μM.

It is possible to prepare nanowires of NPG by first electrodepositing the gold and silver alloy into a template such as anodized aluminum disks which presents a regular array of pores of controllable diameter, with 100–200 nm being typical, of and up to microns in length, in aluminum oxide [41, 83, 84]. To establish the needed electrical contact for electrodeposition, a thin conducting metal layer is deposited on one side of the membrane. The template may be dissolved in KOH and then the nanowires dealloyed in acid to produce NPG nanowires. Using anodized aluminum, electrodeposition from $HAuCl_4$ containing solution can produce a nanostructure similar to NPG constituted of a loose but cohesive packing of nanoparticles [85]. The dissolution of the template left an NPG array on a gold film support that could be used as an electrode. Glucose oxidase was immobilized by physisorption at 4 °C for 24 h and found to perform effectively as a glucose biosensor based upon amperometric detection of H_2O_2. NPG left in the alumina template has been proposed for use as a flow-through composite membrane [84].

1.3 Acetyl-cholinesterase

Acetylcholinesterase is of high interest in sensors for detection of organophosphorous compounds such as pesticides or various nerve agents. The enzyme was immobilized onto NPG [40] by physisorption onto samples of differing average pore size (50–100 nm versus 200 nm) prepared by treating macroscopic plates of a 10 carat gold foil (41.8 atomic% gold) of 250 μm thickness in nitric acid for either 24 h or 72 h. Cross-sectional SEM images confirmed that the nanoporous structure was present throughout the interior of an NPG free-standing plate of this thickness. Longer immersion in strong acid promotes annealing towards larger pore sizes. Enzyme kinetics was studied using the Ellman assay with acetylthiocholine as the substrate. A comparison was made of NPG samples studied either as intact plates of dimension 2.0 mm × 2.0 mm × 0.25 mm or crushed into fragments. The value of K_m was found to increase from 0.08 mM in solution to 0.26 mM for the enzyme on the 50–100 nm pore size NPG and to 0.15 mM for the enzyme on the 200 nm pore size NPG. It was found that K_m values for intact plates and crushed plates of the 24 h acid-treated NPG were similar, with $K_m = 0.32$ mM for a plate crushed before immobilization and $K_m = 0.28$ mM for a plate crushed after immobilization. Enzyme loading, as judged by the values of V_{max}, was significantly larger for both crushed samples. In the case of the sample crushed prior to adsorption, crushing facilitated increased enzyme adsorption while crushing after adsorption enhanced enzyme accessibility to the substrate. K_m values were observed to increase over 6 days of testing, with a greater increase observed for the enzyme on the 200 nm NPG to 2.1 mM. The enzyme loading was determined by analyzing the reduction in activity of the supernatant compared to the initial solution and assuming unchanged kinetic parameters in the solution. Submonolayer coverages of enzyme were found, dependent on the concentration of the bulk enzyme solution. Using measured V_{max} values and estimated enzyme loading, it was determined that k_{cat} for acetylcholinesterase was strongly reduced from the observed value in solution upon adsorption onto NPG.

NPG modified with multi-walled carbon nanotubes (MWCNT) was used for the immobilization of acetylcholinesterase to create a sensor for organophosphates using the pesticide malathion as a test substrate [67]. In this case, the NPG was prepared by a repetitive alloying/dealloying process applied to a gold electrode by electrochemical cycling in $ZnCl_2$ dissolved in benzyl alcohol (25 cycles between 1.8 V and –0.8 V (vs. Zn)). MWCNTs were shortened by reflux in HNO_3 and carboxylic acid groups were also introduced at each end. The NPG was then surface modified with cysteamine and exposed to a solution of MWCNTs. After immobilization of the MWCNTs, the NPG was exposed to a solution of acetylcholinesterase by drop casting and drying. Action of the enzyme on the substrate acetylthiocholine produces thiocholine whose oxidation at the Au surface can be detected amperometrically. The enzyme electrode in the presence of this substrate generated a large catalytic current peak near 913 mV (vs. SCE). Inhibition of the peak current by malathion gave a detection limit of 0.5 ng/ml and a linear range from 1 to 500 ng/ml.

1.4 Other Enzymes

More recently, investigation of a wider range of enzymes on NPG for various applications has begun to be reported. Immobilization of lipase on NPG was successfully used to make an electrochemical sensor for triglycerides [68]. The enzyme was physisorbed for 72 h on an electrode prepared by attachment of acid dealloyed 12 carat Au leaf onto a glassy carbon electrode. Substrates of tributyrin, olive oil, and human serum samples were tested. The hydrolysis of the triglycerides released protons whose reduction was detected by electron transfer through gold to glassy carbon. Use of the electrode gave values for the amount of triglycerides in serum samples in good agreement with a standard lab analysis.

Lignin peroxidase was immobilized on NPG of 40–50 nm average pore diameter by physisorption [70]. The immobilized enzyme lost about 20 % of its activity in one hour at 45 °C compared to free enzyme which lost almost all its activity. The immobilized enzyme was found to be effective at decolorization of dye molecules including fuchsine, rhodamine B, and pyrogallol red, meant to demonstrate that the immobilized enzyme could be useful for degrading aromatic pollutants in water. The immobilized enzyme retained 95 % of its activity when stored at 4 °C for a month.

The enzyme xylanase, useful for hemicellulose conversion in industrial processes, was immobilized on NPG (formed by dealloying 25 μm thick $Au_{42}Ag_{58}$ alloy foil in concentrated HNO_3) by physisorption (10 mg/ml xylanase in 0.2 M pH 6.5 phosphate buffer for 38 h) [71]. X-ray photoelectron spectroscopy (XPS) data indicated a role for formation of Au–S bonds involving cysteine residues. The activity of the immobilized enzyme was studied using xylan as the substrate and determining the amount of xylose produced by stopping the reaction with 3,5-dintrosalicylic acid, which reacts with reducing sugars, and measuring the absorbance at 550 nm. Immobilization increased K_m from 0.12 to 0.27 mM, and decreased k_{cat} from 4024 min^{-1} to 3539 min^{-1}. The amount of immobilized enzyme was determined by analysis of supernatant solution using the Bradford assay.

For the dual applications of removal of Pb^{2+} and degradation of toxic di(ethylhexyl)phthalate from drinking water, the enzyme cutinase was immobilized on NPG (average pore size 35.8 nm, specific surface area of 11.5 m^2/g by BET analysis) that was first surface modified with poylethyleneimine (20,000 average molecular weight) [72]. The NPG was surface modified with lipoic acid and then treated with chloroacetylchlorine in pyridine/chloroform to activate the surface followed by exposure to the polymer in DMF. Cutinase was then physically adsorbed and its activity confirmed by hydrolysis of p-nitrophenyl butyrate by observing the p-nitrophenol product by its absorbance at 405 nm. The enzyme and polymer loaded NPG was effective at removing Pb^{2+} from drinking water and at degrading di(ethylhexyl)phthalate by hydrolysis to 1,3 isobenzofurandione, a nontoxic by-product.

NPG prepared by dealloying 25 μm thick alloy foil of composition Au$_{22}$Ag$_{78}$ was used to immobilized lipase, catalase, and horseradish peroxidase by physisorption [69]. The enzyme loading was determined by analyzing the supernatant and collected washings of the enzyme-loaded NPG using the Bradford assay. The NPG prepared had an average pore size of 35 nm and a specific surface area of 14 m^2/g. Enzyme activity was studied using p-nitrophenyl palmitate as the substrate for lipase and pyrogallol as the substrate for horseradish peroxidase. The activity of catalase was assayed indirectly by stopping the reaction with ammonium molybdate which forms a complex with hydrogen peroxide. Conversion of soybean oil to biodiesel was demonstrated using the immobilized lipase. XPS data gave evidence for Au–S and Au–N interaction which could result in strong attachment of the enzymes to the Au surface after simple physisorption.

1.5 Antibody–Enzyme and Lectin–Enzyme Conjugates on NPG

NPG has also become of interest for immobilization of antibodies or antibody–enzyme conjugates for applications in immunoassays. NPG has been used as a support for an immunoassay for prostate specific antigen (PSA) based on the immobilization of a monoclonal antibody conjugated to the enzyme alkaline phosphatase [4]. NPG was prepared as a ~10 μm thick coating on a gold wire by electrodeposition of a gold + silver alloy of 20 atomic% gold followed by dealloying in nitric acid. The antibody–enzyme conjugate was linked by the EDC coupling reaction to a self-assembled monolayer of lipoic acid formed on the NPG. This immunoassay was based on the principle of inhibition of enzyme activity upon antigen binding with enzyme activity assessed using p-nitrophenyl phosphate as the substrate and spectrophotometric detection of p-nitrophenolate product at 410 nm. The assay was found to respond linearly to PSA up to 20 ng/ml with a detection limit of 0.1 ng/ml. In a prior related study [80, 81], gold-coated nylon membranes were used to develop a colorimetric ELISA for human chorionic gonadotropin (hCG).

NPG prepared by dealloying in HNO_3 Au–Ag alloy that was electrodeposited onto 0.2 mm diameter Au wire to a thickness of ~10 μm on gold wire electrodes was surface modified with SAMs of lipoic acid and used to create electrochemical biosensors for PSA and CEA [73]. Conjugates of IgG monoclonal antibodies with alkaline phosphatase were covalently linked to the lipoic acid SAMs by EDC/NHS coupling. The production of p-aminophenol product from p-aminophenylphosphate substrate was determined using the square wave voltammetry peak current for oxidation of p-aminophenol to p-quinoneimine near 0.2 V (vs. Ag|AgCl). Binding of the PSA or CEA antigen resulted in a reduction in the peak current that was proportional to the antigen concentration. The linear range for PSA response extended up to 30 ng/ml and for CEA extended up to 10 ng/ml. The detection limit for PSA was 0.75 ng/ml and was 0.015 ng/ml for CEA.

Similarly prepared NPG electrodes were used to immobilize conjugates of lectin Concanavalin A and alkaline phosphatase onto lipoic acid SAMs [74]. Using square-wave voltammetry and p-aminophenylphosphate substrate, response to binding of the glycoproteins transferrin or IgG was measured through the decrease in the peak current associated with substrate conversion. Competitive assays were achieved using immobilized glycoprotein and competition between glycoprotein and lectin–enzyme conjugate in solution for binding to the surface.

1.6 Selected Redox Proteins on NPG

The large protein complex photosystem I responsible for photosynthesis in green plants was immobilized on NPG gold leaf electrodes [78]. These electrodes were prepared by floating 12 carat gold leaf onto nitric acid from a glass slide and then transferring it back onto the glass slide and letting it float off again onto distilled water. The gold leaf is only ~100 nm thick and must be handled with care. After dealloying, it was transferred to a Si substrate modified with a 125 nm gold film prepared by thermal evaporation and modified with a self-assembled monolayer of 1,6-hexanedithiol. The exposed thiol group served to bond the NPG piece onto the gold substrate. The NPG pieces were also transferred to gold slides modified with a self-assembled monolayer of mercaptopropyltrimethoxysilane to which they would bond. The NPG was then modified with aminoethanethiol which was then reacted with terephtaldialdehyde. The surface of exposed aldehydes was used to react with lysines of photosystem I to bind it to the surface. Photoelectrochemical measurements confirmed an increase in photocurrent of 3–7 times that observed for a similarly modified flat gold surface. The pore size of the NPG used was mostly 50–100 nm, large enough to accommodate the photosystem I complex of dimensions ~10 nm × 14 nm. The methods used of floating gold leaf onto acid and retrieving it by transfer dipping onto glass slides or silicon are related to methods for manipulating these very thin gold leaf materials introduced by Ding [14].

A material resembling NPG was prepared by creating a positively charged surface by first adsorbing the cationic polymer poly(diallyldimethylammonium) chloride onto a glass slide [82]. Gold nanoparticles of 5 nm diameter were then adsorbed by electrostatic attraction followed by 1,5-pentanedithiol and adsorption of 10 nm silver nanoparticles. After alternation of adsorption of 1,5-pentanedithiol and gold and silver nanoparticles for multiple cycles, the silver nanoparticles were oxidized by addition of $HAuCl_4$ according to the reaction: $AuCl_4^-$ (aq) + 3 Ag(s) → Au (s) + 3 Ag^+ (aq) + 4 Cl^- (aq) in the presence of 3 M NaCl. The resulting structure displays a morphology that resembles ligaments with a particulate appearance. Cytochrome c was adsorbed at 4 °C for 30 min. Direct electrochemistry of cytochrome c was observed and the adsorbed enzyme was effective as an amperometric sensor for H_2O_2.

Hemoglobin immobilized on a highly porous Au was found to display direct electron transfer in cyclic voltammograms that was not observed on a flat Au electrode [54]. The protein was immobilized by adsorption from an aqueous solution that also contained chitosan. The immobilized hemoglobin showed catalytic reduction of H_2O_2 and was highly effective as a biosensor with a detection limit of 0.02 μM and a linear range from 0.05 to 200 μM. The highly porous Au was formed by heating a HCl etched Ti plate in an autoclave with a solution of $HAuCl_4$ and formaldehyde followed by calcining at 250 °C under argon. SEM showed the surface to have a porous open structure of interconnected nanoparticles of average size 22 nm forming pores up to hundreds of nm in size.

A material closely related to NPG which presents gold in a microporous form has been prepared by first forming ordered lattices of 500 nm diameter colloidal latex particles by depositing a small volume of the particle suspension onto a Pt disk and allowing the water to evaporate [82]. After placing the disk in a commercial gold-plating solution, a negative potential is applied. The latex particles are dissolved out using tetrahydrofuran leaving a fairly regular honeycomb like gold structure, but with all of the spherical cavities interconnected by openings between them. The presence of the interconnections between the hollow spherical regions was later confirmed by focused ion beam tomography [83]. The surface area of this "macroporous" gold electrode was estimated by the charge passed during gold oxide stripping. The thickness of the gold microstructure is controlled by the amount of charge passed during the electrodeposition step and is specified as an odd integer number of half sphere layers since the outermost layer represents gold deposited only up to half of the depth of the outermost colloidal particle. The macroporous gold electrode was modified by strong physisorption of a nitrofluorenone (TNF) mediator bearing two nitrile groups that could interact with the gold surfaces [46]. The formation of a monolayer of the mediator was concluded from cyclic voltammetry, and the mediator was quite active towards catalyzing the oxidation of NADH to NAD+. More precise control over the number of layers in the colloidal template has been reported using the Langmuir–Blodgett technique which results in fewer defects in the template than obtained by the evaporation technique [47, 48]. The resulting macroporous gold electrode was used to adsorb the TNF mediator and then the cofactor NAD+ via Ca^{2+} bridges. The enzyme glucose dehydrogenase and its substrate glucose were then added into the solution. An increase in the catalytic current arising from the reoxidation of NADH formed in the enzymatic oxidation of glucose back to NAD+ was observed. In an additional [48], the enzyme was truly immobilized either by cross-linking with glutaraldehyde or by entrapment in an "electrodeposition paint" [86, 87] which is a polymer that becomes insoluble under the influence of sudden decrease in pH brought on by a potential pulse which oxidizes water produces protons near the electrode surface.

Studies of enzyme immobilization have used gold nanowires prepared by electrodeposition into templates such as anodized aluminum followed by dissolution of the template to release the nanowires [49, 50]. Electrodeposition strategies in templates have also been used to produce gold nanotubes [51]. Gold nanowires of 250 nm diameter and 10 μm length were dispersed into a glucose oxidase solution and the enzyme adsorbed overnight at 4 °C [49]. The enzyme modified nanowires were then dispersed into a solution of chitosan and the resulting dispersion cast onto a glassy carbon electrode surface. Covalent attachment of glucose oxidase to gold nanowires modified by self-assembled monolayers of mercaptopropionic acid treated with 2 mM EDC and 2 mM NHS in pH 3.5 buffer was used to create an amperometric glucose sensor [50]. Covalent attachment of the enzyme to a cystamine self-assembled monolayer using glutaraldehyde followed by reaction with enzyme produced a sensor also, but the attachment to mercaptopropionic acid gave a higher response to glucose. The resulting electrode was effective as a glucose sensor using amperometric detection of H_2O_2. An array of gold nanotubes was produced by electroless deposition of gold onto a track-etched polycarbonate template having pores of diameter 460 nm and thickness 20 μm [51]. The template was dissolved using dichloromethane. The enzyme horseradish peroxidase was covalently immobilized by two strategies using different self-assembled monolayers. Conjugation to mono-layers of 2-mercaptoethylamine on the nanotube assembly was achieved by first activating the surface with glutaraldehyde followed by reaction with the enzyme in phosphate buffer. Conjugation to monolayers of mercaptopropionic acid was accomplished by first treating the surface with 2 mM EDC and 5 mM NHS in a pH 3.5 buffer to produced NHS ester groups that were then allowed to react with the enzyme in phosphate buffer. The resulting structure performed with high selectivity as an amperometric sensor for H_2O_2.

A recent study reported the first demonstration of direct entrapment of an enzyme (acid phosphatase) inside nanostructured silver or gold prepared by reducing metal ions directly in the presence of the enzyme [88]. Reduction was achieved by reaction with zinc powder followed by filtering of the precipitate. The enzyme entrapped inside the metallic precipitate could be pressed into a coin shape. Micrographs revealed an NPG like morphology for the silver material showing ligaments but a more bushy structure for the gold material. The enzyme was found to be active and its stability was improved by entrapment.

2   Materials

The materials referred to below are primarily those for experiments reported for enzyme immobilization on NPG obtained by dealloying procedures.

2.1   Solutions and Other Chemicals

1. Electrodeposition of gold-silver alloy solution: 50 mM potassium dicyanoaurate (KAu(CN)$_2$, Sigma-Aldrich) dissolved in 0.1 M sodium carbonate (Na$_2$CO$_3$) (Fisher Scientific, Certified), 50 mM potassium dicyanoargentate (KAg(CN)$_2$, Sigma-Aldrich) dissolved in 0.1 M sodium carbonate (Na$_2$CO$_3$). Combine volumes of these two solutions in the ratio required to achieve the desired Ag–Au ratio in solution which is expected to correspond closely to the ratio obtained in the electrodeposited alloy prepared using the procedure in section 3.2.

2. Thioctic acid: 100 mM α-lipoic acid in ethanol.

3. Mercaptopropionic acid: 2.0 mM in ethanol.

4. Cysteamine: 2.0 mM in ethanol.

5. Hexanedithiol: 20 mM hexanedithiol in ethanol.

6. Mercaptopropyltrimethoxysilane: 5.0 mM in hexane (*see* **Note 1**).

7. EDC: 5 mM N-ethyl-N′-(3-dimethylaminopropyl)carbodiimide (EDC) in acetonitrile.

8. EDC/NHS in buffer: 2 mM EDC and 5 mM *N*-hydroxysuccinimide (NHS) in 50 mM MES (*N*-morpholinoethanesulfonic acid) buffer (pH 3.5).

9. EDC/NHS in water: 75 mM EDC and 15 mM *N*-hydroxysuccinimide (NHS) in water.

10. Glutaraldehyde: 25 % in water.

11. MES buffer: *N*-morpholinoethanesulfonic acid (MES) (pH 3.5, 50 mM).

12. Phosphate buffer: 0.1 M phosphate buffer (pH 7.0).

13. Phosphate-citrate buffer: 0.1 M phosphate-citric acid buffer (pH 6.8).

14. BCA assay kit and or Bradford reagent.

2.2 Materials Specifically for NPG Preparation

1. For preparation of NPG free-standing plate: 10 carat white gold sheet (41.8 atomic percent gold, available in various thicknesses, Hoover and Strong, Richmond, VA).

2. Wire substrate for deposition of NPG precursor alloy: gold wire, 99.9 %, 0.008″ diameter (Electron Microscopy Sciences, PA).

3. For preparation of ultrathin floating NPG: Gold leaf product (9–12 carat) from Sepp Leaf products (New York) such as Monarch brand 12 carat white gold leaf (also available from: www.fineartstore.com); Graphite roller for transferring to and from surface of nitric acid solution and water, or alternately a glass microscope slide may be used. If a graphite roller is used for transfer, then the nitric acid should be placed in a flat glass tray of width slightly greater than that of the roller. If the glass slide is to be used, then a beaker of sufficient depth is adequate.

4. Gold-coated slides for transfer of NPG leaf onto an adhesive support: Platypus Technologies offers gold-coated microscope slides, coverslips, mica, or silicon (Madison, WI); Asylum Research Gold 200C, Gold 500C slides (Santa Barbara, CA). Other suppliers may offer similar products. Either titanium or chromium adhesion layers should be suitable for this purpose.

2.3 Enzymes

1. Enzymes discussed above as having been immobilized on NPG and related materials: glucose oxidase, laccase, horseradish peroxidase, and alkaline phosphatase are commercially available; photosystem I must be extracted from spinach leaves per ref. [79].

2. Alkaline phosphatase conjugated to a monoclonal antibody against prostate specific actigen (Fitzgerald Industries, Concord, MA) was prepared using the alkaline phosphatase labeling kit LK12-10 available from Dojindo Molecular Technologies Inc. (Rockville, MD) and following the instructions included with the kit, as noted in ref. [4]. The kit may be used to label other antibodies as well.

2.4 Instrumentation for NPG Preparation by Electrodeposition

The most accessible routes to obtaining NPG are by dealloying a gold + silver alloy formed by electrodeposition onto a substrate to which it will adhere (such as a gold surface or adhesion layer), or by starting with a commercially available low carat gold alloy piece (9–12 carat, sheet, wire, etc..) cut to the desired dimensions and treating it with strong acid. Methods based on sputter-coating or vacuum deposition to create NPG films on substrates are not described here since they require expensive and generally specially constructed vacuum apparatus. A potentiostat such as Ametek Princeton Applied Research model 2273 or 2263 (Oak Ridge, TN) or equivalent such as CH Instruments 600D series (Austin, TX) capable of chronoamperometry is suitable for electrodeposition. A standard three-electrode electrochemical cell with a silver–silver chloride reference electrode and Pt wire counter-electrode may be used (CH Instruments, Austin, TX).

It will be essential to have access to a scanning electron microscopy (SEM) with sufficient resolution to obtain images of the NPG produced to confirm the porous structure and to characterize pore size. Sample handling is straightforward since gold is an excellent sample for SEM contrast. No preparation other than thorough rinsing and drying is needed prior to imaging. A low-voltage field emission SEM such as JSM 6320F (JEOL Inc., Peabody, MA) is optimal. If available, an accessory for energy dispersive X-ray analysis (EDAX) can be used to confirm the presence of only gold after dealloying.

3 Methods

3.1 Preparation of Free-Standing Nanoporous Gold Plate

1. Macroscopic free-standing pieces of NPG may be prepared by cutting a piece of the alloy foil (or wire) of the desired dimensions and immersing in concentrated nitric acid for at least 24 h. This should be carried out in a covered beaker or dish in a fume hood.

2. After dealloying, the material will have a brownish bronze color. The material will no longer be ductile but can be handled using tweezers. Upon removal from acid, the NPG should be thoroughly rinsed with high purity water by multiple cycles of soaking and removal from water to remove residual acid and dissolved nitrate salts (*see* **Note 2**).

3. The material should be dried under vacuum, especially if it is to be used in a solvent other than water (*see* **Note 3**).

3.2 Preparation of NPG Coating on a Gold Wire or Standard Flat Gold Electrode

1. An NPG electrode can be prepared by electrodepositing a gold and silver alloy onto a conductive substrate such as a gold wire electrode or a flat gold electrode from a solution prepared by combining solutions of 50 mM $KAu(CN)_2$ or 50 mM $KAg(CN)_2$ in 0.10 M Na_2CO_3 in the desired ratio; solutions

combined in a 1:5 to 2:3 volume ratio will yield alloys that will generate NPG after acid treatment.

2. Using a standard three electrode cell and a potentiostat, application of a potential of −1.2 V for 10 min will produce a gold + silver alloy of approximately 10 μm thickness [4, 41]. The composition of the alloy should be very close to the mole ratio of gold and silver cyanide ions in solution.

3. Dealloying may then be achieved by immersion in concentrated nitric acid for 2 h followed by rinsing. Dealloying may also be achieved electrochemically by applying a potential near or above the critical potential for dealloying which is 0.90 V (vs. NHE, normal hydrogen electrode) for an alloy of 20 atomic% gold, 80 atomic% silver in 0.1 M perchloric acid [17, 18].

3.3 Preparation of NPG Free-Floating Piece from Gold Leaf Material

1. Carefully lay a piece of gold leaf (*see* **Note 4**) on the surface of a graphite roller or microscope slide. Handle the gold leaf carefully using flat head tweezers.

2. Partially fill a rectangular glass tray with concentrated nitric acid.

3. Roll the graphite cylinder partly immersed in the acid along the tray such that the NPG floats off onto the acid.

4. The dealloying should take place quickly due to the thin nature of the sample; 15 min should be sufficient [14].

5. Using the graphite roller, transfer the dealloyed sheet which is now the NPG sample back onto the roller.

6. The NPG sheet may be rolled off to float on water for rinsing and then recovered. If it is wished to transfer the NPG onto a substrate to which it will adhere, this can be a glass slide functionalized with mercaptopropyltrimethoxysilane monolayer prepared by exposure to a 5 mM solution in hexane for 1 h at 60 °C. A glass slide or silicon wafer onto which a chromium adhesion layer followed by a gold layer have been thermally evaporated with subsequent modification with a self-assembled monolayer of hexanedithiol (20 mM in ethanol for 24 h) may also be used. The preparation of these gold-coated slides in-house requires use of a thermal evaporator or a sputter coater and is highly nontrivial. Commercially prepared gold-coated substrates may be purchased directly, as described above.

3.4 Enzyme Immobilization by Physisorption

Immobilization of an enzyme by simple physisorption is typically accomplished by immersing the NPG overnight at 4 °C in a solution of the enzyme in a buffer appropriate for the enzyme followed by removal and rinsing.

3.5 Enzyme Immobilization by Electrostatic Immobilization

The isoelectric point of the enzyme must be known [12]. A positively charged surface may be used to electrostatically immobilize an enzyme from a buffer of pH > pI where the enzyme is negatively charge. A negatively charged surface may be used to immobilize an enzyme at pH < pI where the enzyme is of positively charged. Immobilization of laccase (pI = 3.5) has been accomplished from 50 mM buffers of pH = 4–8. The NPG was first treated with methylene blue, a cationic dye molecule that will form an adsorbed monolayer on NPG. The methylene blue was adsorbed from a 1.0 mM solution for 24 h at 4 °C. The reported amount of enzyme adsorbed by this method increases with pH. This method results in enzyme loading that is much less stable to leaching than physisorption or covalent coupling.

3.6 Enzyme Immobilization by Covalent Coupling Reaction Using EDC/NHS

1. Immerse the NPG sample in the 100 mM thiooctic acid solution for 15–24 h at room temperature. Alternatively, a 1.0 mM solution of lipoic acid in water. A solution of mercaptopropionic acid or similar carboxylic acid terminated alkanethiol in ethanol (2 mM) may also be used.

2. Remove the NPG sample and rinse thoroughly with water. It also is useful to soak the sample in water to allow unbound molecules to diffuse out of the pores of the sample.

3. Immerse the sample into an aqueous solution of 75 mM EDC and 15 mM NHS for 10 h at room temperature. Variations on the EDC/NHS activation of the carboxylic acid groups of self-assembled monolayers have been reported, including application of 2 mM EDC and 5 mM NHS in an MES buffer (pH 3.5) for 2 h to achieve activation.

4. Remove the sample and rinse thoroughly with water.

5. Immerse the sample into a solution of the desired enzyme (1 mg/ml typical concentration) in phosphate buffer (pH 7.0) for 24 h at 4 °C.

6. Remove sample and rinse thoroughly with buffer.

3.7 Enzyme Immobilization by Covalent Coupling Reaction Using EDC in Acetonitrile [4, 81]

1. Immerse the NPG sample in 100 mM lipoic acid in ethanol for 15 h at room temperature.

2. Remove the sample and rinse thoroughly with ethanol and then dry under vacuum.

3. Immerse the sample in 5 mM EDC in acetonitrile for 5 h at room temperature.

4. Remove the sample and rinse thoroughly with water.

5. Immediately immerse the sample in a solution of the enzyme (1 mg/ml typical concentration) in 0.1 M phosphate buffer at pH 7.0 for 24 h at 4 °C.

3.8 Enzyme Immobilization by Covalent Coupling Reaction Using Glutaraldehyde

1. Immerse the NPG sample into a 2 mM solution of mercaptoethylamine in ethanol for 15–24 h at room temperature.

2. Remove the sample and rinse thoroughly with ethanol.

3. Dilute a portion of a commercial glutaraldehyde solution (25 % in water) with 0.1 M phosphate buffer of pH 7.0 by a factor of 100.

4. Immerse the sample into the diluted glutaraldehyde solution for 2 h at room temperature.

5. Remove the sample and rinse thoroughly with water.

6. Immerse the sample into a solution of the enzyme (1 mg/ml typical concentration) for 15–24 h at 4 °C.

7. Remove the sample and rinse thoroughly with water.

3.9 Characterization of Enzyme Loading Using Protein Concentration Assay

1. The well-known BCA assay or Bradford assay is recommended [79]. Compatibility of the chosen assay method with the protein should be checked; for example, the BCA assay gives anomalous results for cysteine rich proteins and the Bradford assay does for arginine rich proteins. Assay kits are available along with instructions for their use. Access to a UV–visible spectrometer is required such as a Cary 50 (Varian Products, Palo Alto, CA).

2. Conduct an assay of the protein concentration in the initial enzyme solution (*See* **Note 5**). The enzyme should be prepared in a buffer solution that is compatible with the selected assay method [79] in both the nature and concentrations of its components.

3. Carry out the enzyme immobilization procedure using the prepared enzyme solution.

4. Retain the supernatant solution remaining after enzyme immobilization.

5. Rinse the NPG sample with buffer and determine the volume collected.

6. Conduct an assay of the protein concentration in the supernatant solution and that collected by rinsing. The amount of protein lost upon immobilization can be determined in micrograms from the concentration difference and known volumes. Using the mass of the NPG sample, enzyme loading in terms of mass of enzyme per gram of NPG may be determined. In order to determine a surface coverage of enzyme in the NPG, data on the surface area of the prepared NPG in m^2/g is required.

4   Notes

1. The use of ethanol from plastic bottles should be avoided for self-assembled monolayer formation. Plasticizers are likely to contaminate the gold surface and interfere with monolayer formation. The use of HPLC grade solvents in general is recommended for solutions to be used for self-assembled monolayer formation.

2. NPG may be stored indefinitely under pure solvent or buffer. However, should the buffer become contaminated, SEM

imaging may show micron sized dark cylindrical objects scattered across the surface that are most likely bacteria.

3. NPG has a strong tendency to retain water and this water is not easily replaced by other solvents. Drying the NPG under vacuum is the best way to remove the trapped water prior to use of NPG in another solvent.

4. Gold leaf is a fragile material, especially after dealloying and is best handled in the absence of any air currents.

5. The volume and concentration of enzyme solution and amount of NPG added must be considered carefully if it is desired to determine enzyme surface loading using a protein concentration assay. It is desirable that the amount of enzyme loaded into the NPG deplete the enzyme solution by a significant and readily measured amount in the vicinity of 10–40%. This consideration is aided by an estimate of the gold surface area in the sample found by using its mass multiplied by its surface area in m^2/g.

References

1. Lavrik NV, Tipple CA, Sepaniak MJ, Datskos PG (2001) Gold nano-structures for transduction of biomolecular interactions into micrometer scale movements. Biomed Microdev 3:35–44

2. Tipple CA, Lavrik NV, Culha M, Headrick J, Datskos P, Sepaniak MJ (2002) Nanostructured microcantilevers with functionalized cyclodextrin receptor phases: self-assembled monolayers and vapor-deposited films. Anal Chem 74:3118–3126

3. Hieda M, Garcia R, Dixon M, Daniel T, Allara D, Chan MHW (2004) Ultrasensitive quartz crystal microbalance with porous gold electrodes. Appl Phys Lett 84:628–630

4. Shluga OV, Zhou D, Demchenko AV, Stine KJ (2008) Detection of free prostate specific antigen on a nanoporous gold platform. Analyst 133:319–322

5. Wittstock A, Biener J, Bäumer M (2010) Nanoporous gold: a new material for catalytic and sensor applications. Phys Chem Chem Phys 12:12919–12930

6. Pornsuriyasak P, Ranade S, Li A, Parlato MC, Sims C, Shulga OV, Stine KJ, Demchenko AV (2009) STICS: surface-tethered iterative carbohydrate synthesis. Chem Commun 1834–1835

7. Ganesh NV, Fujikawa K, Tan Y-H, Stine KJ, Demchenko AV (2013) Surface-tethered iterative carbohydrate synthesis (STICS): a spacer study. J Org Chem 78:6849–6857

8. Yin H, Zhou C, Xu C, Liu P, Xu X, Ding Y (2008) Aerobic oxidation of D-glucose on support-free nanoporous gold. J Phys Chem C 112:9673–9678

9. Nagle LC, Rohan JF (2011) Nanoporous gold anode catalyst for direct borohydride fuel cell. Intl J Hydrogen Energ 103:319–326

10. Zhang J, Liu P, Ma H, Ding Y (2007) Nanostructured porous gold for methanol electro-oxidation. J Phys Chem C 111:10382–10388

11. Qiu H, Xu C, Huang X, Ding Y, Qu Y, Gao P (2008) Adsorption of Laccase on the surface of nanoporous gold and the direct electron transfer between them. J Phys Chem C 112:14781–14785

12. Qiu H, Xu C, Huang X, Ding Y, Qu Y, Gao P (2009) Immobilization of laccase on nanoporous gold: comparative studies on the immobilization strategies and the particle size effects. J Phys Chem C 113:2521–2525

13. Fujita T, Qian L-H, Inoke K, Erlebacher J, Chen M-W (2008) Three-dimensional morphology of nanoporous gold. Appl Phys Lett 92:251902/1-3

14. Ding Y, Kim Y, Erlebacher J (2004) Nanoporous gold leaf: "ancient technology" advanced materials. Adv Mater 16:1897–1900

15. Dixon MC, Daniel TA, Hieda M, Smilgies DM, Chan MHW, Allara DL (2007) Preparation, structure, and optical properties of nanoporous gold thin films. Langmuir 23:2414–2422

16. Wittstock A, Biener J, Erlebacher J, Baumer M (eds) (2012) Nanoporous gold: from an ancient technology to a high-tech material. Royal Society of Chemistry

17. Sieradzki K, Dimitrov N, Movrin D, McCall C, Vasiljevic N, Erlebacher J (2002) The dealloying critical potential. J Electrochem Soc 149:B370–B377

18. Dursun A, Pugh DV, Corocoran SG (2005) Probing the dealloying critical potential. morphological characterization and stead-state current behavior. J Electrochem Soc 152:B65–B72

19. Senior NA, Newman RC (2006) Synthesis of tough nanoporous metals by controlled electrolytic dealloying. Nanotechnology 17:2311–2316

20. Snyder J, Livi K, Erlebacher J (2008) Dealloying silver/gold alloys in neutral silver nitrate solution: porosity evolution, surface composition, and surface oxides. J Electrochem Soc 155:C464–C473

21. Detsi E, De Jong E, Zinchenko A, Vukovic Z, Vukovic I, Punzhin S, Loos K, ten Brinke G, De Raedt HA, Onck PR, De Hosson JTM (2011) On the specific surface area of nanoporous material. Acta Mater 59:7488–7497

22. Detsi E, van de Schootbrugge M, Punzhin S, Onck PR, De Hosson JTM. On tuning the morphology of nanoporous gold. Scripta Materialia 64:319–322

23. Qian LH, Chen MW (2007) Ultrafine nanoporous gold by low-temperature dealloying and kinetics of nanopore formation. Appl Phys Lett 91:083105–3

24. Lu X, Balk TJ, Spolenak R, Arzt E (2007) Dealloying of Au-Ag thin films with a composition gradient: influence on morphology of nanoporous Au. Thin Solid Films 515:7122–7126

25. Huang J-F, Sun I-W (2006) Fabrication and surface functionalization of nanoporous gold by electrochemical alloying/dealloying of Au-Zn in an ionic liquid, and the self-assembly of L-cysteine monolayers. Adv Funct Mater 15:989–994

26. Xu Y, Ke X, Yu C, Liu S, Zhao J, Cui G, Higgins D, Chen Z, Li Q, Wu G (2014) A strategy for fabricating nanoporous gold films through chemical dealloying of electrochemically deposited Au-Sn alloys. IOP Nanotech 24:445602

27. Chen S, Chu Y, Zheng J, Li Z (2009) Study on the two dealloying modes in the electrooxidation of Au–Sn alloys by in situ Raman spectroscopy. Electrochim Acta 54:1102–1108

28. El-Mel A-A, Boukli-Hacene F, Molina-Luna L, Bouts N, Chauvin A, Thiry D, Gautron E, Gautier N, Tessier P-Y (2015) Unusual dealloying effect in gold/copper alloy thin films: the role of defects and column boundaries in the formation of nanoporous gold. ACS Appl Mater Interf

29. Morrish R, Dorame K, Muscat AJ (2011) Formation of nanoporous Au by dealloying AuCu thin films in HNO_3. Scr Mater 64:856–859

30. Renner FU, Stierle A, Dosch H, Kolb DM, Lee TL, Zegenhagen J (2008) In situ X-ray diffraction study of the initial dealloying and passivation of Cu_3Au (111) during anodic dissolution. Phys Rev B 77:235433

31. Pareek A, Borodin S, Bashir A, Ankah GN, Keil P, Eckstein GA, Rohwerder M, Stratmann M, Gründer Y, Renner FU. Initiation and inhibition of dealloying of single crystalline Cu_3Au (111) surfaces. J Am Chem Soc 133:18264–18271

32. Zhang Z, Wang Y, Wang Y, Wang X, Qi Z, Ji H, Zhao C (2010) Formation of ultrafine nanoporous gold related to surface diffusion of gold adatoms during dealloying of Al_2Au in an alkaline solution. Scr Mater 62:137–140

33. Zhang Z, Zhang C, Gao Y, Frenzel J, Suna J, Eggeler G (2012) Dealloying strategy to fabricate ultrafine nanoporous gold-based alloys with high structural stability and tunable magnetic properties. Cryst Eng Commun 14:8292–8300

34. Zhang Q, Wang X, Qia Z, Wang Y, Zhang Z (2009) A benign route to fabricate nanoporous gold through electrochemical dealloying of Al–Au alloys in a neutral solution. Electrochim Acta 54:6190–6198

35. Zhang Z, Wang Y, Qi Z, Lin J, Bian X (2009) Nanoporous gold ribbons with bimodal channel size distributions by chemical dealloying of Al-Au alloys. J Phys Chem C 113:1308–1314

36. Cortie MB, Maaroof AI, Smith GB (2005) Electrochemical capacitance of mesoporous gold. Gold Bull 38:14–22

37. Detsi E, Punzhin S, Rao J, Onck PR, De Hosson JTM. Enhanced strain in functional nanoporous gold with a dual microscopic length scale structure. ACS Nano 6:3734–3744

38. Lang XY, Guo H, Chen LY, Kudo A, Yu JS, Zhang W, Inoue A, Chen MW (2010) Novel nanoporous Au-Pd alloy with high catalytic activity and excellent electrochemical stability. J Phys Chem C 114:2600–2603

39. Snyder J, Asanithi P, Dalton AB, Erlebacher J (2008) Stabilized nanoporous metals by dealloying ternary alloy precursors. Adv Mater 20:4883–4886

40. Shulga OV, Jefferson K, Khan AR, D'Souza VT, Liu J, Demchenko AV, Stine KJ (2007) Preparation and characterization of porous gold as a platform for immobilization of acetylcholine esterase. Chem Mater 19:3902–3911

41. Ji C, Searson PC (2003) Synthesis and characterization of nanoporous gold nanowires. J Phys Chem B 107:4494–4499

42. Tan YH, Davis JA, Fujikawa K, Ganesh NV, Demchenko AV, Stine KJ (2012) Surface area and pore size characteristics of nanoporous gold subjected to thermal, mechanical, or chemical modifications studied using gas adsorption isotherms, cyclic voltammetry, and scanning electron microscopy. J Mater Chem 22:6733–6745

43. Erlebacher J, Aziz MJ, Karma A, Dimitrov N, Sieradzki K (2001) Evolution of nanoporosity in dealloying. Nature 410:450–453

44. Erlbacher J, Sieradzki K (2003) Pattern formation during dealloying. Scripta Mater 991–996

45. Li R, Sieradzki K (1992) Ductile-brittle transition in random porous Au. Phys Rev Lett 68:1168–1171

46. Ben-Ali S, Cook DA, Evans SAG, Thienpoint A, Bartlett PN, Kuhn A (2003) Electrocatalysis with monolayer modified highly organized macroporous electrodes. Elect Commun 5:747–751

47. Szamocki R, Reculusa S, Ravaine S, Bartlett PN, Kuhn A, Hempelmann R (2006) Tailored mesostructuring and biofunctionalization of gold for increased electroactivity. Angew Chem Intl Ed 45:1317–1321

48. Szamocki R, Velichko A, Holzapfel C, Mucklich F, Ravaine S, Garrigue P, Sojic N, Hempelmann R, Kuhn A (2007) Macroporous ultramicroelectrodes for improved electroanalytical measurements. Anal Chem 79:533–539

49. Csumá A, Curulli A, Zane D, Kaciulis S, Padeletti G (2007) Feasibility of enzyme biosensors based on gold nanowires. Mater Sci Eng C 27:1158–1161

50. Lu Y, Yang M, Qu F, Shen G, Yu R (2007) Enzyme-functionalized gold nanowires for the fabrication of biosensors. Bioelectrochemistry 71:211–216

51. Delvaux M, Walcarius A, Demoustier-Champagne S (2004) Electrocatalytic H_2O_2 amperometric detection using gold nanotube electrode ensembles. Anal Chim Acta 525:221–230

52. Bonroy K, Friedt J-M, Frederix F, Laureyn W, Langerock S, Campitelli A, Sara M, Borghs G, Goddeeris B, Declerck P (2004) Realization and characterization of porous gold for increased protein coverage on acoustic sensors. Anal Chem 76:4299–4306

53. Jia F, Yu C, Ai Z, Zhang L (2007) Fabrication of nanoporous gold film electrodes with ultrahigh surface area and electrochemical activity. Chem Mater 19:3648–3653

54. Kafi AKM, Ahmadalinezhad A, Wang J, Thomas DF, Chen A (2010) Direct growth of nanoporous gold and its application in electrochemical biosensing. Biosens Bioelectron 25:2458–2463

55. Hou C, Yang D, Liang B, Liu A (2014) Enhanced performance of a glucose/O_2 biofuel cell assembled with laccase-covalently immobilized three-dimensional macroporous gold film-based biocathode and bacterial surface displayed glucose dehydrogenase-based bioanode. Anal Chem 86:6057–6063

56. du Toit H, Di Lorenzo M (2014) Glucose oxidase directly immobilized onto highly porous gold electrodes for sensing and fuel cell applications. Electrochim Acta 138:86–92

57. Cherevko S, Chung C-H (2011) Direct electrodeposition of nanoporous gold with controlled multimodal pore size distribution. Electrochem Commun 13:16–19

58. Tan YH, Schallom JR, Ganesh NV, Fujikawa K, Demchenko AV, Stine KJ (2011) Characterization of protein immobilization on nanoporous gold using atomic force microscopy and scanning electron microscopy. Nanoscale 3:3395–3407

59. Hakamada M, Takahashi M, Mabuchi M (2012) Enzyme electrodes stabilized by monolayer-modified nanoporous Au for biofuel cells. Gold Bull 45:9–15

60. Salaj-Kosla U, Poller S, Schuhmann W, Shleev S, Magner E (2013) Direct electron transfer of Trametes hirsute laccase adsorbed at unmodified nanoporous gold electrodes. Bioelectrochemistry 91:15–20

61. Hakamada M, Mabuchi M (2014) Stabilization and decomposition of organic matters by nanoporous metals. Proc Mater Sci 4:335–340

62. Scanlon MD, Salaj-Kosla U, Belochapkine S, MacAodha D, Leech D, Ding Y, Magner E (2012) Characterization of nanoporous gold electrodes for bioelectrochemical applications. Langmuir 28:2251–2261

63. Salaj-Kosla U, Pöller S, Beyl Y, Scanlon MD, Beloshapkin S, Shleev S, Schuhmann W, Magner E (2012) Direct electron transfer of bilirubin oxidase (*Myrothecium verrucaria*) at an unmodified nanoporous gold biocathode. Electrochem Commun 16:92–95

64. Chen LY, Fujita T, Chen MW (2012) Biofunctionalized nanoporous gold for electrochemical biosensors. Electrochim Acta 67:1–5

65. Xiao X, Ulstrup J, Li H (2014) Nanoporous gold assembly of glucose oxidase for electrochemical biosensing. Electrochim Acta 130:559–567

66. Wu C, Sun H, Li Y, Liu X, Du X, Wang X, Xu P (2015) Biosensor based on glucose oxidase—nanoporous gold co-catalysis for glucose detection. Biosens Bioelectron 66:350–355

67. Ding J, Zhang H, Jia F, Qin W, Du D (2014) Assembly of carbon nanotubes on a nanoporous gold electrode for acetylcholinesterase biosensor design. Sens Act B 199:284–290

68. Wu C, Liu X, Li Y, Du X, Wang X, Xu P (2014) Lipase-nanoporous gold modified electrode for reliable detection of triglycerides. Biosens Bioelectron 53:26–30

69. Wang X, Liu X, Yan X, Zhao P, Ding Y, Xu P (2011) Enzyme-nanoporous gold biocomposite: excellent biocatalyst with improved biocatalytic performance and stability. PLoS One 6:e24207

70. Qiu H, Li Y, Ji G, Zhou G, Huang X, Qu Y, Gao P (2009) Immobilization of lignin peroxidase on nanoporous gold: enzymatic properties and in situ release of H_2O_2 by co-immobilized glucose oxidase. Biores Tech 100:3837–3842

71. Yan X, Wang X, Zhao P, Zhang Y, Xu P, Ding Y (2012) Xylanase immobilized nanoporous gold as a highly active and stable biocatalyst. Micro Mesopor Mater 161:1–6

72. Zhang C, Zeng G, Huang D, Lai C, Huang C, Li N, Xu P, Cheng M, Zhou Y, Tang W, He X (2014) Combined removal of di(2-ethylhexyl) phthalate (DEHP) and Pb(II) by using a cutinase loaded nanoporous gold-polyethyleneimine adsorbent. RSC Adv 4:55511–55518

73. Pandey BP, Demchenko AV, Stine KJ (2012) Nanoporous gold as a solid support for an electrochemical immunoassay for prostate specific antigen (PSA) and carcinoembryonic antigen (CEA). Microchim Acta 179:71–81

74. Pandey B, Bhattarai JB, Pornsuriyasak P, Fujikawa K, Catania R, Demchenko AV, Stine KJ (2014) Square-wave voltammetry assays for glycoproteins on nanoporous gold. J Electroanal Chem 717–718:47–60

75. Minteer SD, Liaw BY, Cooney MJ (2007) Enzyme-based biofuel cells. Curr Opin Biotech 18:228–234

76. Liu Z, Huang L, Zhang L, Ma H, Ding Y (2011) Electrocatalytic oxidation of D-glucose at nanoporous Au and Au–Ag alloy electrodes in alkaline aqueous solutions. Electrochim Acta 54:7286–7293

77. Sotiropoulou S, Vamvakaki V, Chaniotakis NA (2005) Stabilization of enzymes in nanoporous materials for biosensor applications. Biosens Bioelectron 20:1674–1679

78. Ciesielski PN, Scott AM, Faulkner CJ, Berron BJ, Cliffel DE, Jennings GK (2008) Functionalized nanoporous gold leaf for the immobilization of photosystem I. ACS Nano 2:2465–2472

79. Olsen BJ, Markwell J (2007) Assays for the determination of protein concentration. Curr Protocols Protein Sci. Suppl 48, unit 3.4, Wiley Interscience, Malden MA

80. Duan C, Meyerhoff ME (1995) Immobilization of proteins on gold-coated porous membranes via an activated self-assembled monolayer of thioctic acid. Mikrochim Acta 117:195–206

81. Smith AM, Ducey MW, Meyerhoff ME (2000) Nature of immobilized antibody layers linked to thioctic acid treated gold surfaces. Biosens Bioelectron 15:183–192

82. Zhu A, Tian Y, Liu H, Luo Y (2009) Nanoporous gold film encapsulating cytochrome c for the fabrication of a H_2O_2 biosensor. Biomaterials 30:3183–3188

83. Laocharoensuk R, Sattayasamitsathit S, Burdick J, Kanatharana P, Thavarungkul P, Wang J (2007) Shape-tailored porous gold nanowires: from nano barbells to nano step-cones. ACS Nano 1:403–408

84. Liu L, Lee W, Huang Z, Scholz R, Gösele U (2008) Fabrication and characterization of a flow-through nanoporous gold nanowire/AAO composite membrane. Nanotech 19:335604/1-6

85. Zhang X, Li D, Bourgeois L, Wang H, Webley PA (2009) Direct electrodeposition of porous gold nanowire arrays for biosensing applications. ChemPhysChem 10:436–441

86. Kurzawa C, Hengstenberg A, Schuhmann W (2002) Immobilization method for the preparation of biosensors based on pH shift-induced deposition of biomolecule-containing polymer films. Anal Chem 74:355–361

87. Neugebauer S, Stoica L, Guschin D, Schuhmann W (2008) Redox-amplified biosensors based on selective modification of nanopore electrode structures with enzymes entrapped within electrodeposition paints. Microchim Acta 163:33–40

88. Ben-Knaz R, Avnir D (2009) Bioactive enzyme-metal composites: the entrapment of acid phosphatase within gold and silver. Biomaterials 30:1263–1267

Chapter 6

Enzyme Stabilization via Bio-Templated Silicification Reactions

Glenn R. Johnson and Heather R. Luckarift

Abstract

Effective entrapment of enzymes in solid phase materials is critical to their practical application. The entrapment generally stabilizes biological activity compared to soluble molecules and the material simplifies catalyst integration compared to other methods. A silica sol-gel process based upon biological mechanisms of inorganic material formation (biomineralization) supports protein immobilization reactions within minutes. The material has high protein binding capacity and the catalytic activity of the enzyme is retained. We have demonstrated that both oligopeptides and selected proteins will mediate the biomineralization of silica and allow effective co-encapsulation of other proteins present in the reaction mixture. The detailed methods described here provide a simple and effective approach for molecular biologists, biochemists and bioengineers to create stable, solid phase biocatalysts that may be integrated within sensors, synthetic processes, reactive barriers, energy conversion, and other biotechnology concepts.

Key words Lysozyme, Biomineralization, Silica, Sol-gel, Biosensor, Biocatalysis, Enzyme immobilization, Butyrylcholinesterase, Silicification

1 Introduction

There are many well established conventional sol-gel chemistries for encapsulating proteins [1, 2]. The earliest report appeared when Dickey used the materials he designed for specific adsorption of organic dyes to trap catalase, urease, and cytochrome C in silica and each appeared to retain functional conformations [3]. The initial experiments showed promise, but the limitations of classic sol-gel processes for protein encapsulation include harsh alkaline or acidic conditions, denaturation by alcohols, limited porosity and "shrinking" of the matrix as it cures, and relatively long development time (hours to days) for the material to undergo poly condensation and cure to the final product [4]. Since that time, the sol-gel concept has been extensively explored and a series of modifications has enhanced biocompatibility of the process and the finished material [5]. Two significant hurdles still remain using

Shelley D. Minteer (ed.), *Enzyme Stabilization and Immobilization: Methods and Protocols*, Methods in Molecular Biology, vol. 1504, DOI 10.1007/978-1-4939-6499-4_6, © Springer Science+Business Media New York 2017

conventional chemistries; the curing time required for maturation of conventional sol-gels and the marginal binding/trapping of protein in many instances.

Organisms form complexes containing inorganic materials using a process generically referred to as "biomineralization." Examples of the final materials include bone, invertebrate exoskeletons, egg shells, and teeth [6]. Marine diatom cell walls form in a biomineralization process in which anionic polypeptides (silaffins) control precipitation of silica nanospheres from solutions of silicic acid in membrane vesicles within the organism [7, 8]. The process can be reproduced in vitro using synthetic peptides in a rapid reaction to yield uniformly sized (~500 nm) silica spheres [9]. If an additional enzyme is present in the reaction solution, it is effectively trapped within the nanosphere architecture [10, 11].

In vitro biomineralization is similar to the sol-gel process, but also holds great advantages for the entrapment of biological molecules. Both methods occur at ambient temperatures and can be adapted to limit co-solvents and contact with acid or base catalysts; aspects that help preserve biological activity. The conventional sol-gels are nearly solid polymers after curing. The silica formed during the biomimetic reaction is a mesoporous material (pore size 2–50 nm) with a comparably open architecture. That openness allows for better diffusion of molecules into biomimetic silica, which will result in high catalytic activities of the immobilized proteins. The biomineralization initially yields silica nanospheres ($\approx$8 nm dia), which aggregate to form a raspberry-like architecture for the final silica form [12]. The speed and ease of the biomimetic reaction is a huge benefit compared to the hours or days of curing time for conventional reactions. In addition, the silica encapsulation reaction can be modified to coat surfaces with films of enzymes or to make particles/powder that contains enzymes. Lastly, the weight protein binding capacity for biomimetic silica is significantly greater than any of the conventional methods described to date [13].

2   Materials

All standard laboratory reagents are available from Sigma-Aldrich (St. Louis, MO) or Fisher Scientific (Pittsburgh, PA) unless otherwise stated.

2.1 Silicification Reaction with R5 Peptide

1. Silicification buffer: Mix 520 mL of 0.1 M monobasic potassium phosphate, (13.6 g/L) with 480 mL of 0.1 N NaOH (4 g/L). Confirm that the pH is 8 and store at room temperature. If needed, add drops of 1 M NaOH or 1 M HCl to adjust the pH as required (*see* **Note 1**).

2. Stock solution of 1 M HCl: Add 8.2 mL of concentrated HCl (37.5%) to 100 mL DI water (*see* **Notes 2** and **3**). **Caution: Always add acid to water.** Dilute the 1 M stock (1 mL into

1 L) to give a working concentration of 1 mM HCl which can be stored at room temperature.

3. Stock solution of 1 M tetramethylorthosilicate (TMOS): mix 0.148 mL of (98%) TMOS with 0.852 mL of 1 mM HCl (*see* **Note 4**). **Caution: TMOS is toxic and should be handled with care in a fume hood. Gloves and eyewear should be worn and exposure should be limited.** The TMOS solution is not immediately miscible; so mix well on a vortex until completely homogeneous (*see* **Note 5**).

4. Stock solution of R5 peptide (sequence H$_2$N-SSKKSGSY SGSKGSKRRIL-COOH; New England Peptides, crude purity with no terminal modifications): dissolve peptide in DI water to give a final concentration of 100 mg/mL.

5. Stock solution of butyrylcholinesterase: dissolve butyrylcholinesterase (EC 3.1.1.8) [highly purified lyophilized powder from equine serum containing approximately 50% protein and activity of 1200 units/mg protein] in silicification buffer to give 50 units/mL (based on the nominal units of activity as designated by the supplier). Store at –20 °C in small (e.g., 100 µL) aliquots, for future use.

2.2 Butyrylcholinesterase Activity Assay

1. Stock solution of indophenyl acetate: dissolve 0.375 g of indophenyl acetate in 25 mL of ethanol to give a final concentration of 6.2×10^{-4} M. Dilute 0.1 mL stock solution in 10 mL silicification buffer (pH 8). [The extinction coefficient of indophenyl acetate is 3.98 M^{-1} cm^{-1} at 460 nm and 15.8 M^{-1} cm^{-1} at 270 nm] [14] (*see* **Note 6**).

2.3 Silicification Reaction with Hen Egg White Lysozyme

1. Silicification buffer: as described in Subheading 2.1, **item 1**.

2. Solution of 1 mM HCl: as described in Subheading 2.1, **item 2**.

3. Stock solution of 1 M tetramethylorthosilicate (TMOS): as described in Subheading 2.1, **item 3**.

4. Stock solution of hen egg white lysozyme (hereafter, lysozyme) is dissolved in silicification buffer to give a final concentration of 100 mg/mL.

2.4 Measurement of Protein Concentration

1. Dilution buffer: Add 87 g of K$_2$HPO$_4$ to 0.5 L of DI water to give 1 M dibasic potassium phosphate. Add 68 g of KH$_2$PO$_4$ to 0.5 L of DI water to give 1 M monobasic potassium phosphate. Mix 61.5 mL of the 1 M dibasic with 38.5 mL of the 1 M monobasic and make up to 1 L with DI water to give 100 M potassium phosphate buffer, pH7. Confirm that the pH is 7 and store at room temperature.

2. Using Pierce® bicinchoninic acid (BCA) protein assay kit (Pierce, Thermo Scientific, Rockford, IL), prepare working reagent by adding 50 mL reagent A to 1 mL of green reagent B and mix well (*see* **Note 7**).

Table 1
Preparation of diluted protein standards

Vial	Volume of diluent (see Note 9) (μL)	Volume of lysozyme stock	Final lysozyme concentration (μg/mL)
A	0	300 μL of stock	2000
B	125	375 μL of stock	1500
C	325	325 μL of stock	1000
D	175	175 μL of vial B	750
E	325	325 μL of vial C	500
F	325	325 μL of vial E	250
G	325	325 μL of vial F	125
H	400	100 μL of vial G	25
I	400	0 μL	0 (Blank)

3. Accurately prepare a solution of lysozyme at 2 mg/mL [or use the standard bovine serum albumin provided in the assay kit] (*see* **Note 8**). From the stock solution, prepare a series of dilutions as described in Table 1.

4. Samples for analysis must be diluted in dilution buffer to fit within the working range of the assay kit (20–2000 μg/mL) (*see* **Note 10**).

2.5 Enzymatic Assay of Lysozyme: Liquid Assay

1. Add 0.015 g *Micrococcus lysodeikticus* cells (ATCC 4698; lyophilized cells) and 0.1 g of bovine serum albumin to 100 mL of 100 mM phosphate buffer, (pH 7) (*see* **Note 11**).

2. A standard solution of lysozyme is prepared in dilution buffer at a final concentration of 100 μg/mL.

3. Using the protein concentrations determined from the BCA assay (Subheading 2.4), prepare dilutions of silica samples in dilution buffer to give a final concentration of 100 μg/mL.

2.6 Enzymatic Assay of Lysozyme: Plate Assay

1. *Micrococcus* agar plates: Add 5 g agar (Difco™ granulated agar, BD, Franklin Lakes, NJ) to 500 mL of 100 mM, potassium phosphate buffer (pH 7) and autoclave at 121 °C for 15 min (*see* **Note 12**). After sterilization, allow the agar to cool to ~60 °C and add 0.5 mg/mL *Micrococcus lysodeikticus* cells and mix well (*see* **Note 13**). Pour plates immediately and allow to set and dry (*see* **Note 14**).

2. Sterilize BBL™ blank paper disks, (6 mm diameter; BD) by autoclaving at 121 °C for 15 min.

3   Methods

The following will help outline the speed, ease and effectiveness of the biomimetic silica reaction for construction of solid-phase biocatalysts. In the example procedures, two different template molecules will act as mediators of silica formation. The first template (R5) is derived from the natural silaffin protein from diatoms [7, 8, 15]. In the second example, lysozyme serves as the template [16]. In the second approach, the catalytically active lysozyme is entrapped in the matrix, providing an added functionality to the materials (which is also outlined in the description). Likewise, the silica architecture will still accommodate a co-encapsulated biocatalyst [17]. Other peptides will mediate in vitro silica mineralization [18], and if appropriately selected, also add functionality to the material [19]. The alternate template methods may influence the morphology and intrinsic biochemistry of the material and these factors offer several questions that could be explored further. This method highlights encapsulation of butyrylcholinesterase as the model enzyme. In practice, the researcher will substitute their biocatalyst of interest. Recent work demonstrates the general applicability of the method with several classes of enzyme and invites additional investigation to test its potential [13].

3.1 Silicification Reaction with R5 Peptide

1. In a 2 mL microcentrifuge tube (or other suitable container) (*see* **Note 15**), mix 0.8 mL of silicification buffer with 0.1 mL R5 peptide stock (100 mg/mL).

2. Add 0.1 mL TMOS (1 M) to the microcentrifuge tube and shake briefly (10 s) on a vortex mixer. Silica formation occurs rapidly and will be evident by the transparent solution changing to a white opaque suspension (Fig. 1). Leave for 5 min.

3. Place the microcentrifuge tube (now containing the silica suspension) directly into a centrifuge and spin at $14,000 \times g$ for 5 min to collect the silica precipitate (*see* **Note 16**).

4. Remove the supernatant with a plastic or glass pipette, transfer to a microcentrifuge tube and store on ice for protein analysis (Subheading 3.4).

5. Add 1 mL of silicification buffer to the silica pellet and mix well until the silica is resuspended (*see* **Note 17**).

6. Centrifuge again at $14,000 \times g$ for 5 min and repeat this wash **step 3** times in total. Save the wash fractions after each step and store on ice for analysis.

7. Resuspend the silica pellet in 0.8 mL of silicification buffer and use for subsequent enzyme assays.

3.2 Encapsulation of Butyrylcholinesterase in Peptide-Mediated Silica

1. In a 2 mL microcentrifuge tube (or other suitable container) (*see* **Note 15**), mix 0.8 mL of butyrylcholinesterase (50 units/mL in silicification buffer) with 0.1 mL R5 peptide stock (100 mg/mL).

2. Follow **steps 2–6** as described in Subheading 3.1, **step 1**. Save the supernatant and wash fractions after each step and store on ice for

Fig. 1 Time-lapse photography of silica formation

analysis. These fractions contain the residual butyrylcholinesterase and can be used to measure the efficiency of encapsulation.

3. Resuspend the silica pellet in 0.8 mL of silicification buffer and use for subsequent enzyme assays (*see* **Note 18**).

3.3 Determination of Butyrylcho-linesterase Enzyme Activity

In this chapter, the encapsulation of butyrylcholinesterase is described with associated enzyme assays to exemplify the encapsulation technique and to demonstrate the modifications to conventional enzyme assays that may be required when measuring the activity of the silica-encapsulated enzyme. Protein determination of the encapsulated enzyme, for example, cannot be measured directly due to the interfering presence of the template (R5) peptide, which will also react with the protein determination reagent. The techniques described are adaptable, but enzyme activity determination must be replaced with relevant assays.

1. For the soluble enzyme samples of butyrylcholinesterase; add 100 μL of (50 units/mL) butyrylcholinesterase to 0.9 mL silicification buffer in a 1 mL cuvette and mix well. Add 10 μL of 0.2 mM indophenyl acetate and mix well by repeated pipetting. Monitor the absorbance at 630 nm after 2 min and record the value.

2. By varying the concentration of butyrylcholinesterase (units) in **step 1**, a calibration curve can be generated using the absorbance reading at 630 nm (at 2 min) vs. butyrylcholinesterase concentration, as shown in Fig. 2 (*see* **Note 19**).

3. For the silica-encapsulated butyrylcholinesterase, dilute the sample (assuming a maximum theoretical concentration of

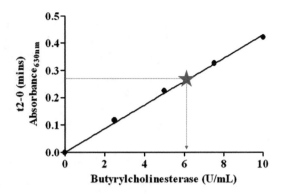

Fig. 2 Calibration plot for butyrylcholinesterase-catalyzed conversion of indophe-nyl acetate. Absorbance at 630 nm after 2 min of reaction time is plotted against butyrylcholinesterase concentration (units/mL). The absorbance of the silica-encapsulated butyrylcholinesterase under the same reaction conditions (unknown; designated by *red star*) can be read from the calibration curve to determine the units of butyrylcholinesterase in the silica-encapsulated sample

encapsulated butyrylcholinesterase of 100% (i.e., 50 units/mL in this example) within the range of the calibration data from **step 2**. Following the assay the efficiency of enzyme encapsula-tion can be indirectly determined using calibration curve above.

4. Add 0.1 mL of silica-butyrylcholinesterase to 0.9 mL silifica-tion buffer in a 1 mL cuvette and mix well. Add 10 μL of 0.2 mM indophenyl acetate and mix well by pipetting up and down. Monitor the absorbance at 630 nm after 2 min. Stop the reaction after 2 min incubation by removing the particles by centrifugation (10 s at $14,000 \times g$) and transfer the superna-tant to a cuvette. Read the absorbance of the supernatant at 630 nm and record the value.

5. Apply the value in **step 4** to the calibration curve to determine the units of butyrylcholinesterase in the sample and allow for any dilutions (*see* **Note 20**).

3.4 Precipitation of Silica Using Lysozyme as Template

The precipitation of silica using lysozyme as a template is an example of self- encapsulation in which, the template molecule is also the catalyst used for further application. Silica formation using lysozyme is described with associated enzyme assays to exemplify the encapsulation technique and to demonstrate the modifications to conventional enzyme assays that may be required when measuring the activity of the silica-encapsulated enzyme. In this case, protein determination of the encapsulated enzyme can be measured directly. As noted previously, the techniques described are adaptable but enzyme activity deter-mination must be replaced with assays relevant to the enzyme of choice.

1. In a 2 mL microcentrifuge tube (or other suitable container) (*see* **Note 15**), mix 0.8 mL of silicification buffer with 0.1 mL lysozyme (100 mg/mL).

2. Add 0.1 mL (1 M) TMOS to the solution in the microcentrifuge tube and shake briefly (10 s) on a vortex mixer. Ass before, silica formation occurs rapidly and will be evident by the transparent solution changing to a white opaque suspension. Leave for 5 min.

3. Place the microcentrifuge tube (containing the now precipitated silica) directly into a centrifuge and spin at $14,000 \times g$ for 5 min to collect the silica precipitate (*see* **Note 16**).

4. Remove the supernatant (which contains the residual lysozyme) with a plastic or glass pipette, transfer to a microcentrifuge tube and store on ice for protein analysis.

5. Add 1 mL of silicification buffer to the silica pellet and vortex until the silica is resuspended (*see* **Note 17**).

6. Centrifuge again at $14,000 \times g$ for 5 min and repeat this wash **step 3** times in total. Save the wash fractions after each step and store on ice for analysis.

7. Resuspend the silica pellet in 1 mL of silicification buffer and use for subsequent enzyme assays (*see* **Note 21**).

3.5 Mass Balance Calculation of Lysozyme Encapsulation

1. The supernatant and wash fractions from the silicification reaction contain unreacted lysozyme and can be used to determine the efficiency of encapsulation. Because the concentration of lysozyme in these samples is unknown, a series of dilutions should be prepared for protein determination. The maximum concentration is 10 mg/mL of lysozyme (assuming 0 % encapsulation); therefore an undiluted (neat) sample and dilutions of 1:10, 1:25, 1:50, and 1:100 in dilution buffer are recommended.

2. The silica pellet contains encapsulated lysozyme and can be measured by the BCA protein assay. Dilute the sample as described above.

3. Add 25 µL of each standard and each of the sample dilutions (in triplicate) into the wells of a microtiter plate.

4. Add 200 µL of the BCA working reagent to each well and shake for approximately 30 s to aid in mixing.

5. Cover the plate and incubate at 37 °C for 30 min (*see* **Note 22**).

6. Allow the plate to cool to room temperature and record the absorbance of each well at 562 nm (*see* **Note 23**). Prepare a standard curve by plotting the blank-corrected absorbance vs. concentration (µg/mL protein) for each standard and analyze by linear regression. The protein concentration of each unknown sample can be extrapolated from the graph. Any

Table 2
Typical mass balance of silica formation catalyzed by lysozyme as determined by protein concentration

Fraction	Concentration of lysozyme (mg/mL)	Volume (mL)	Total lysozyme (mg)	Mass balance (% of total)
Initial reaction	9.98 ± 0.23	1	9.98	100
Silica pellet	8.45 ± 0.97	1	8.45	84
Supernatant	0.49 ± 0.08	1	0.49	4.9
Wash	0.33 ± 0.09	3	0.87	8.7

dilutions which give readings outside the working range of the assay (20–2000 µg/mL) should be discarded (*see* **Note 24**).

7. Calculate the lysozyme concentration in the silica pellet, reaction supernatant and wash fractions. Remember to allow for any dilutions and factor in the volume of the wash fractions.

The mass balance of a typical lysozyme mediated silica reaction is shown in Table 2.

3.6 Determination of Lysozyme Enzyme Activity: Liquid Assay

1. Based on the protein concentration from the BCA protein assay (Subheading 3.4), prepare a dilution of soluble lysozyme and a dilution of silica-encapsulated lysozyme at a final concentration of 100 µg/mL.

2. Pipette 1 mL of *Micrococcus lysodeikticus* cell suspension into a plastic cuvette and place into a spectrophotometer at room temperature.

3. Monitor the absorbance at 450 nm until a stable baseline is established.

4. Add 10 µL of standard (soluble lysozyme) or sample (silica-encapsulated lysozyme) at 100 µg/mL to the *Micrococcus* cells and mix well by repeated pipetting. Monitor the absorbance over time and calculate the change in absorbance over time from the slope of the graph (*see* **Note 25**).

5. Compare the activity of each sample to determine the efficiency of encapsulation.

3.7 Determination of Lysozyme Enzyme Activity: Plate Assay

1. Bring *Micrococcus* agar plates to room temperature.

2. Place a sterile paper filter disk in the center of each plate using sterile forceps.

3. Based on the protein concentration from the BCA protein assay (Subheading 3.4), prepare a dilution of soluble lysozyme and silica-encapsulated lysozyme at a final concentration of 100 µg/mL.

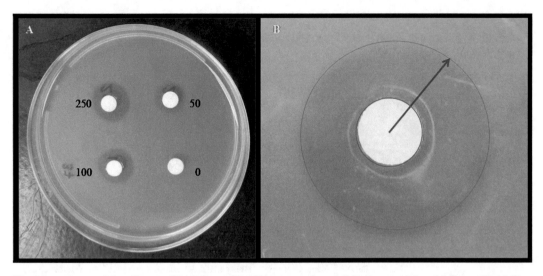

Fig. 3 Lysozyme catalyzed zone of inhibition with *Micrococcus lysodeikticus* agar plates. (**a**) Zone of inhibition from silica-encapsulated lysozyme shown at a range of concentrations (μg/mL). (**b**) Photographically enlarged zone of inhibition showing calculation of lytic radius from the center of the filter disk to the outer edge of the zone of inhibition (inhibition circles are *highlighted* for clarity)

4. Pipette 10 μL of each standard and sample to a filter disk and incubate at 37 °C for 24 h.

5. Measure the clear zone surrounding caused by lysozyme lysis of the *Micrococcus* cells (Fig. 3).

4 Notes

1. 0.1 M Potassium or sodium phosphate buffer (pH 8) can be used interchangeably.

2. Concentrated HCl as a 37.5 % solution ($\delta = 1.189$ g/mL) is 12.2 M, a 1 M HCl solution contains 82 mL/L.

3. Throughout, DI water refers to deionized water (Barnstead Nanopure water purification system, Thermo Scientific). The resulting DI water is ultra filtered purified water with a resistivity of ~18 MΩ.

4. The TMOS stock must be prepared fresh daily. It will polymerize over a period of 24 h.

5. TMOS (tetramethylorthosilicate) can be replaced with TEOS (tetraethylorthosilicate) as a less toxic reagent. The use of TEOS will still allow for silica formation but the versatility of using TEOS to make silica for encapsulation of proteins has not been fully characterized in our laboratory.

6. Indophenyl acetate in buffer at pH 8 will gradually change from yellow to green over a period of approximately 24 h. Make fresh dilutions daily.

7. The working reagent is stable for several days in the refrigerator but should be discarded if the solution begins to lose its bright green color. Adjusting the volume of working reagent to prepare sufficient volume for the assays required is advised.

8. The Pierce BCA assay protocol uses bovine serum albumin provided with the kit. Lysozyme at 2 mg/mL can be used as standard in this assay and provides a more accurate measure of protein concentration in the reactions in which lysozyme precipitates silica formation and eliminates protein–protein variation for the assay method.

9. Use the same buffer for diluent as used for sample preparation.

10. If an approximate estimation of protein concentration is not known, a series of dilutions should be prepared.

11. The absorbance at 450 nm of this suspension should be between 0.6 and 0.7. The *Micrococcus lysodeikticus* cell suspension can be stored in the refrigerator for several days. The cells will settle upon standing, and can be suspended by shaking.

12. Always vent vessels prior to autoclaving to avoid pressure build up. 500 mL of agar solution will make ~20 standard agar plates. The volume can be adjusted accordingly.

13. Dispersion of the *Micrococcus* cells is best achieved by weighing out 250 mg cells and transferring to a small amount of sterile water (~5 mL), mix the cell suspension with the cooled agar and swirl gently to mix but avoid extensive bubble formation.

14. If the plates are not used immediately after drying, store in an airtight bag in the refrigerator. If left at room temperature, the *Micrococcus* cells will grow; the plates will become turbid and should be discarded.

15. Using a microcentrifuge tube allows for immediate centrifugation in the next step.

16. The speed and timing of the silica collection step is not critical and can be adapted depending on the type of centrifuge available. Spin until the silica pellets and the supernatant is transparent.

17. If the pellet is densely packed, the silica can be resuspended using a glass tissue homogenizer.

18. Resuspending the pellet in 0.8 mL ensures that the enzyme has not been diluted from the starting concentration (50 units/mL) and provides for direct comparison with the soluble enzyme at the same concentration.

19. Butyrylcholinesterase hydrolyzes the yellow indophenyl acetate to a blue reaction product (4-(4-hydroxy-phenylimino)-cyclohexa-2,5-dienone). The extinction coefficient of the product was

determined from this method to be 8100 M^{-1} cm^{-1} based on complete conversion to product by butyrylcholinesterase.

20. To create a silica blank for spectrophotometer readings, use the silica as prepared in Subheading 3.1, **step 1** (in the absence of butyrylcholinesterase) and dilute to give the same absorbance spectrum as the sample. Add 10 μL of 0.2 mM indophenyl acetate and use this as a reagent blank. This control eliminates any deviation in the absorbance readings associated with non-specific adsorption of indophenyl acetate to the silica particles.

21. Note that the starting concentration was 10 mg of lysozyme in 1 mL, resuspending the resulting silica particles in 1 mL of buffer provides a lysozyme dilution of 10 mg/mL (assuming 100 % encapsulation).

22. If an incubator is not available, the reaction can be incubated at room temperature for 2 h before reading.

23. Water can be used as a blank to set the spectrophotometer baseline. For silica samples, the blank should contain a comparable amount of diluted silica to allow for any variation in absorbance due to the silica particles.

24. A second set of sample dilutions should be prepared if analysis suggests protein concentration is outside the working limits of the assay.

25. Lysozyme is extremely catalytically active. High enzyme concentrations in this assay cause misleading results. Lysozyme should be diluted until the activity can be measured as a linear slope over at least 2 min.

Acknowledgements

The research related to the presented methods was supported by the Air Force Research Laboratory Materials Science Directorate, the Air Force Office of Scientific Research (Program Managers, Walt Kozumbo and Jennifer Gresham), and the Joint Science and Technology Office-Defense Threat Reduction Agency (Program Managers, Jennifer Becker and Stephen Lee).

References

1. Gill I, Ballesteros A (1998) Encapsulation of biologicals within silicate, siloxane, and hybrid sol gel polymers: An efficient and generic approach. J Am Chem Soc 120:8587–8598

2. Mansur H, Orefice R, Vasconcelos W, Lobato Z, Machado L (2005) Biomaterial with chemically engineered surface for protein immobilization. J Mater Sci Mater Med 16:333–340

3. Dickey FH (1955) Specific adsorption. J Phys Chem 59:695–707

4. Braun S, Rappoport S, Zusman R, Avnir D, Ottolenghi M (1990) Biochemically active sol-gel glasses: the trapping of enzymes. Mater Lett 10:1–5

5. Brennan J (2007) Biofriendly sol-gel processing for the entrapment of soluble and

membrane-bound proteins:toward novel sold-phase assays for high-throughput screening. Acc Chem Res 40:827–835

6. Baeuerlein E (2007) Handbook of biomineralization: biological aspects and structure formation, vol 2. Wiley-VCH Verlag GmbH & Co. KGaA, Weinheim

7. Kröger N, Deutzmann R, Sumper M (1999) Polycationic peptides from diatom biosilica that direct silica nanosphere formation. Science 286:1129–1132

8. Kröger N, Lorenz S, Brunner E, Sumper M (2002) Self-assembly of highly phosphorylated silaffins and their function in biosilica morphogenesis. Science 298:584–586

9. Brott LL, Naik RR, Pikas DJ, Kirkpatrick SM, Tomlin DW, Whitlock PW, Clarson SJ, Stone MO (2001) Ultrafast holographic nanopatterning of biocatalytically formed silica. Nature 413:291–293

10. Luckarift HR, Spain JC, Naik RR, Stone MO (2004) Enzyme immobilization in a biomimetic silica support. Nat Biotechnol 22:211

11. Naik RR, Tomczak MM, Luckarift HR, Spain JC, Stone MO (2004) Entrapment of enzymes and nanoparticles using biomimetically synthesized silica. Chem Commun. 1684–1685

12. Cardoso MB, Luckarift HR, Urban VS, O'Neil H, Johnson GR (2010) Protein localization in silica nanosphere derived via biomimetic mineralization. Adv Funct Mater. 3031–3038

13. Betancor L, Luckarift HR (2008) Bioinspired enzyme encapsulation for biocatalysis. Trends Biotechnol 26:566–572

14. Sigma-Aldrich. Personal communication

15. Kroger N, Deutzmann R, Sumper M (2001) Silica-precipitating peptides from diatoms. The chemical structure of silaffin-A from *Cylindrotheca fusiformis*. J Biol Chem 276:26066–26070

16. Luckarift HR, Dickerson MB, Sandhage KH, Spain JC (2006) Rapid, room-temperature synthesis of antibacterial bionanocomposites of lysozyme with amorphous silica or titania. Small 2:640–643

17. Ivnitski D, Artyushkova K, Rincon RA, Atanassov P, Luckarift HR, Johnson GR (2008) Entrapment of enzymes and carbon nanotubes in biologically synthesized silica: glucose oxidase-catalyzed direct electron transfer. Small 4:357–364

18. Naik RR, Brott LL, Clarson SJ, Stone MO (2002) Silica-precipitating peptides isolated from a combinatorial phage display peptide library. J Nanosci Nanotechnol 2:95–100

19. Eby DM, Farrington KE, Johnson GR (2008) Synthesis of bioinorganic antimicrobial peptide nanoparticles with potential therapeutic properties. Biomacromolecules 9:2487–2494

Chapter 7

Covalent Immobilization of Enzymes on Eupergit® Supports: Effect of the Immobilization Protocol

Zorica D. Knežević-Jugović, Sanja Ž. Grbavčić, Jelena R. Jovanović, Andrea B. Stefanović, Dejan I. Bezbradica, Dušan Ž. Mijin, and Mirjana G. Antov

Abstract

A selection of best combination of adequate immobilization support and efficient immobilization method is still a key requirement for successful application of immobilized enzymes on an industrial level. Eupergit® supports exhibit good mechanical and chemical properties and allow establishment of satisfactory hydrodynamic regime in enzyme reactors. This is advantageous for their wide application in enzyme immobilization after finding the most favorable immobilization method. Methods for enzyme immobilization that have been previously reported as efficient considering the obtained activity of immobilized enzyme are presented: direct binding to polymers via their epoxy groups, binding to polymers via a spacer made from ethylene diamine/glutaraldehyde, and coupling the periodate-oxidized sugar moieties of the enzymes to the polymer beads. The modification of the conventionally immobilized enzyme with ethylenediamine via the carbodiimide route seems to be a powerful tool to improve its stability and catalytic activity.

Key words Covalent immobilization, Glycoenzymes, Periodate oxidation, Eupergit C, Epoxy groups, Aminated support, Enzyme chemical amination, Multipoint covalent attachment, Carbodiimide chemistry

1 Introduction

Being applied as catalysts, enzymes have certain advantages over chemical catalysts, including mild reaction conditions, a high catalytic activity, high specificity, and economic viability. To fully exploit the technical and economic advantages of enzymes, it is recommended to use them in an immobilized state to reduce cost and the poor stability of the soluble form [1, 2]. The immobilization also facilitates the development of continuous bioprocesses, enhances enzyme properties, such as stability and activity under extreme conditions of temperature, pH, or in the presence of organic solvents, and provides more flexibility with the enzyme–substrate contact by using various reactor configurations [3–5].

Shelley D. Minteer (ed.), *Enzyme Stabilization and Immobilization: Methods and Protocols*, Methods in Molecular Biology, vol. 1504, DOI 10.1007/978-1-4939-6499-4_7, © Springer Science+Business Media New York 2017

For successful development of an immobilized enzyme system, the enzyme support is generally considered as the most important component contributing to the performance of the reactor system. Eupergit® (made by copolymerization of N,N'-methylene-bis-(methacrylamide), glycidyl methacrylate, allyl glycidyl ether, and methacrylamide) deserves special attention because it exhibits good chemical, mechanical and other properties such as low swelling tendency in common solvents, ability to handle high flow rates in column, and an excellent performance in stirred batch reactors [6]. The presence of epoxy groups in the Eupergit® provides a binding site for enzymes, which makes its use as a matrix for immobilizing several enzymes very attractive [7–9]. Immobilization of enzymes on Eupergit® is mostly achieved by direct enzyme coupling to the polymer via epoxy groups. Since the sugar residues in glycoenzymes are often not required for their activities, a more efficient strategy, which includes binding to polymers via activated carbohydrate moieties, seems to be very promising [10–12]. This type of binding results in favorable orientation of the immobilized enzyme since its active site region moves to further distance from the support, facilitating interaction with a substrate. The method can be applicable for a wide variety of lipases and other glycoproteins but can also be applied for nonglycosylated enzymes immobilization on epoxy supports. In this case, the alternative procedure is to chemically modify the enzyme to prepare a synthetic glycoconjugate with much better possibilities of yielding a more suitable covalent attachment during immobilization. This modified enzyme will then be covalently bound to epoxy-supports in a similar way that the other glycoenzymes [13].

This chapter deals with the covalent immobilization of enzymes on Eupergit® by three different protocols: (1) a direct enzyme binding to polymers via their epoxy groups, (2) enzyme binding to polymers via a spacer made from ethylene diamine/glutaraldehyde, and (3) coupling the periodate-oxidized sugar moieties of the enzymes to the polymer beads. An approach to first standard method consisting of the post-immobilization chemical modification of enzymes with ethylenediamine via the carbodiimide chemistry to increase the intensity of the enzyme–support multipoint covalent attachment is also presented. The concept involves the introduction of additional amino groups into enzyme molecules allowing the completion of the immobilization process, i.e., the formation of multiple cooperative chemical bonds which is considered as a major contributing factor to the high operational stability of enzymes bound to Eupergit® supports [14, 15].

2   Materials

1. *Candida rugosa* lipase type VII (Sigma, St. Louis, USA).
2. Eupergit® C and 250 L (Rohm GmbH & Co., Degussa, Darmstadt, Germany).

2.1   The Preparation of Amino-Eupergit®

1. 2% (v/v) ethylenediamine (EDA, 1,2-diaminoethane), pH 10.0.
2. Sodium borohydride.

2.2   Immobilization Procedure I

1. Immobilization buffer: 1.0 M potassium phosphate buffer, pH 7.0 (prepare 1 M potassium dihydrogen phosphate and adjust pH to 7.0 using 1 M dipotassium hydrogen phosphate). Dilute a portion of this buffer to provide 0.1 M phosphate buffer, pH 7.0 (*see* **Notes 1** and **2**).
2. Blocking buffer: 0.2 M solution of ethanolamine (2-aminoethanol) in 0.2 M phosphate buffer, pH 8.0.
3. Washing solution: 0.1 M potassium phosphate buffer, pH 7.0 and the same buffer containing 1 M NaCl.

2.3   Immobilization Procedure II

1. Activation buffer: 0.1 M phosphate buffer, pH 8.0. Prepare 0.1 M potassium dihydrogen phosphate and adjust pH to 8.0 using 0.1 M dipotassium hydrogen phosphate.
2. Glutaraldehyde solution, store at 0 °C. Dilute glutaraldehyde aqueous solution 25% (w/v) in 0.1 M potassium phosphate buffer, pH 8.0 to produce 2.5% (w/v).
3. Immobilization buffer: 0.1 M phosphate buffer, pH 7.0. Prepare 0.1 M potassium dihydrogen phosphate and adjust pH to 7.0 using 0.1 M dipotassium hydrogen phosphate.
4. Washing solutions: 0.1 M phosphate buffer, pH 7.0 and the same buffer containing 1 M NaCl.
5. Rotary stirrer.

2.4   Immobilization Procedure III

2.4.1   Oxidation of Enzyme

1. Oxidation buffer: 0.1 M sodium acetate buffer, pH 5.0 (*see* **Note 3**). Prepare 0.1 M sodium acetate and adjust pH to 5.0 using 0.1 M acetic acid. Dilute a portion of this buffer to provide 0.05 M sodium acetate buffer, pH 5.0.
2. 0.1 M sodium *meta*-periodate ($NaIO_4$) in 0.1 M sodium acetate buffer, pH 5.0.
3. Blocking buffer: 0.01 M aqueous ethylene glycol solution.
4. Dialyzing buffer: 0.05 M sodium acetate buffer, pH 5.0.

2.4.2   Coupling of Oxidized Enzyme to Amino-Eupergit® Support

1. Immobilization buffer: 0.1 M sodium acetate buffer, pH 5.0.
2. Reducing reagent: sodium borohydride, 4 mg/mL (freshly prepared).

3. Dialyzing buffer: 0.05 M sodium acetate buffer, pH 5.0.

4. Washing buffer: 0.1 M phosphate buffer, pH 7.0 and the same buffer containing 1 M NaCl.

5. Storage buffer: 0.1 M phosphate buffer containing 0.05% sodium azide.

2.5 Immobilization Procedure IV

2.5.1 Immobilization of Enzyme

The enzyme was directly immobilized on Eupergit® as described in Subheading 2.2 and submitted to the amination process.

2.5.2 Post-immobilization Chemical Amination of Enzyme

1. Amination reagent: 1 M ethylenediamine (EDA) at pH 4.75. EDA solution is adjusted to pH 4.75 with 0.1 M HCl.

2. Activation reagent: a water soluble solid carbodiimide, EDAC (1-ethyl-3-(dimethylamino-propyl) carbodiimide) (Sigma-Aldrich).

3. Tyrosine residues recovering solution: 0.1 M hydroxylamine solution, pH 7. Add a certain amount hydroxylamine HCl to a volumetric flask (e.g., 0.174 g hydroxylamine HCl to a 25 mL volumetric flask) and fill the flask to the mark with deionized water and mix.

2.6 Determination of Lipase Activity

1. Solution A: prepare solution A by dissolving 30 mg of p-nitrophenylpalmitate in 10 mL of isopropanol.

2. Sodium phosphate buffer: prepare 0.1 M sodium dihydrogen phosphate aqueous solution. Adjust pH to 8.0 using 0.1 M disodium hydrogen phosphate aqueous solution.

3. Solution B: prepare solution B by dissolving 0.1 g of gum arabic and 0.4 mL of Triton X-100 in 90 mL of phosphate buffer (pH 8.0) (see Note 4).

4. Substrate solution: Prepare substrate solution by adding solution A to solution B dropwise with constant stirring to obtain an emulsion that remains stable for 2 h (see Note 5).

5. Spectrophotometer for the absorbance reading at 410 nm.

2.7 Determination of Protein Content

1. The reagent solution: dissolve 100 mg Coomassie brilliant blue G250 (see Note 6) in a mixture of 100 mL 85% (w/v) phosphoric acid and 50 mL 95% (v/v) ethanol. After the dye has dissolved, add water to bring the final volume to 1 L (if necessary, the solution can be filtered through Whatman no. 1 paper to remove any insoluble material). This solution is stable for several weeks when stored in the dark at room temperature.

2. Bovine serum albumin (BSA) solution: 1 mg BSA (Sigma-Aldrich)/mL in appropriate buffer (see Note 7).

3. Spectrophotometer for the absorbance reading at 595 nm.

3 Methods

Three different approaches for covalent immobilization of enzyme are presented on Eupergit® C or Eupergit® C 250L: (1) a direct enzyme binding to polymers via their epoxy groups; (2) enzyme binding to polymers via a spacer made from ethylene diamine/glutaraldehyde; (3) glycoenzyme binding to amino-Eupergit through the enzyme carbohydrate moiety previously modified by periodate oxidation. To enhance the stability of the immobilized enzyme, an attempt of variation of the first standard method of enzyme immobilization on epoxy-supports is also presented. The concept involves the introduction of additional amino groups into enzyme molecules allowing the completion of the immobilization process, i.e., the formation of multiple cooperative chemical bonds [15].

3.1 Immobilization via Epoxy Groups

The standard method for enzyme immobilization on Eupergit® supports involves the direct enzyme binding on polymers via epoxy groups as shown in Fig. 1.

1. Dissolve 50 mg of the enzyme in 50 mL of 1.0 M potassium phosphate buffer (1 mg/mL, pH 7.0, *see* **Note 8**). Take a sample for determination of protein content and enzyme activity (0.1 mL).

2. Suspend unmodified Eupergit C or C 250 L (500 mg) in 50 mL of native enzyme solution in 1.0 M potassium phosphate buffer, pH 7.0 and stir the mixture thoroughly to ensure complete mixing.

3. Incubate the reaction mixture at 25 °C for 48 h. (*see* **Note 9**).

4. Treat the immobilized enzyme beads with 0.2 M solution of ethanolamine in phosphate buffer to block remaining active groups.

5. Separate the immobilized enzyme beads by vacuum filtration using a sintered glass filter (*see* **Note 10**).

6. Thoroughly wash the beads on the filter with 100 mL of 0.1 M phosphate buffer (pH 7.0), afterwards three times with 100 mL of the same buffer containing 1 M NaCl. Do not discard supernatant and washing solutions; instead collect the washings separately for determination of non-immobilized protein and enzyme activity to estimate the efficiency of immobilization.

3.2 Preparation of Amino-Eupergit® Supports

The epoxy groups of the Eupergit® can react with small diamine compounds to yield short alkylamine spacer that can be used for subsequent conjugation reaction.

1. Add 30 mL of 2% (v/v) ethylenediamine (EDA) at pH 10.0 to 5 g of the polymer under very gentle stirring.

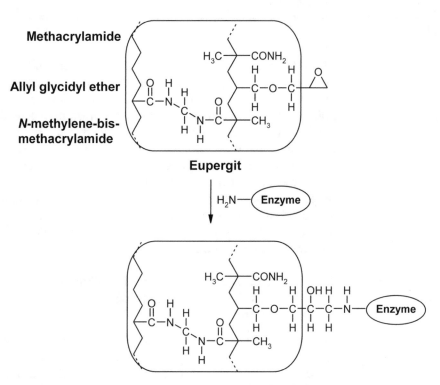

Methacrylamide

Allyl glycidyl ether

N-methylene-bis-methacrylamide

Eupergit

H₂N—(Enzyme)

Fig. 1 Schematic illustration of the enzyme immobilization on Eupergit® via epoxy groups

2. Maintain the mixture under agitation at 100 rpm for 2 h in an ice bath.

3. Add 1.15 g of sodium borohydride and maintain under agitation at 100 rpm for another 2 h.

4. After reduction of the amino-polymer beads, decant the mixture and eliminate the supernatant.

5. Wash the activated particles several times with an excess of distilled water and store at 4 °C (*see* **Note 11**).

3.3 Immobilization of Enzyme by Glutaraldehyde

The immobilization procedure consists of two main steps: polymer pretreatment with 1,2-diaminoethane (EDA) followed by glutaraldehyde activation and enzyme coupling to the activated polymer.

3.3.1 Activation of Eupergit Support

The epoxide groups of the Eupergit may be reacted with small diamine compounds to yield short alkylamine spacer that can be further activated with glutaraldehyde. The obtained active aldehyde derivative of Eupergit® reacts with amino groups on the enzyme to form a Schiff base linkage as shown in Fig. 2 (*see* **Note 12**).

1. Add 5 g of amino-Eupergit in 20 mL of the glutaraldehyde solution buffered at pH 8.0 (0.1 M potassium phosphate buffer) and readjust pH 8.0.

2. Stir the mixture for 2 h at 150 rpm (overhead stirrer or shaker) at room temperature (20–25 °C).

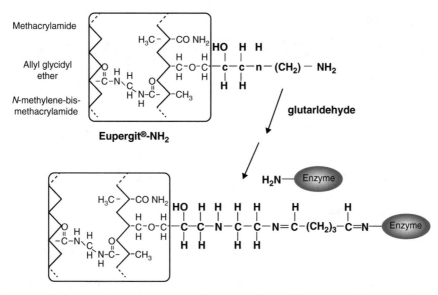

Fig. 2 Schematic illustration of the enzyme immobilization on Eupergit® by glutaraldehyde method

3. Decant the mixture and eliminate the supernatant.

4. Wash the activated particles several times with water and the phosphate buffer.

3.3.2 Enzyme Coupling to the Activated Polymers

1. Suspend the activated polymer particles in 20 mL of native enzyme solution in potassium phosphate buffer, pH 7.0 (e.g., for lipase from *C. rugosa* 100 mg crude lipase g^{-1} carrier).

2. Incubate for a minimum of 20 h at room temperature (20–25 °C) with stirring (≥150 rpm).

3. Recover the immobilized enzyme beads by filtration on a sintered glass filter by suction under vacuum (*see* **Note 10**).

4. Thoroughly wash the beads with 100 mL of 0.1 M phosphate buffer, pH 7.0, then three times with 100 mL of the same buffer containing 1 M NaCl to remove physically adsorbed enzyme from the carrier.

5. Collect the washing solutions separately for protein and enzyme assays.

3.4 Immobilization of Enzyme by Periodate Method

The immobilization procedure consists of three main steps: oxidation of lipase by sodium *meta*-periodate, the polymer pretreatment with 1,2-diaminoethane, and coupling of oxidized enzyme to amino-Eupergit supports (Fig. 3). The oxidation of glycoproteins with sodium *meta*-periodate (e.g., lipase from *Candida rugosa*) cleaves bonds between adjacent carbon atoms that contain hydroxyl groups (*cis*-glycols) of the carbohydrate residues, creating two aldehyde groups. The generated aldehydes then can be used in coupling reactions with amine-activated supports to form Schiff bases that can be stabilized to secondary amine bonds by reduction with sodium borohydride.

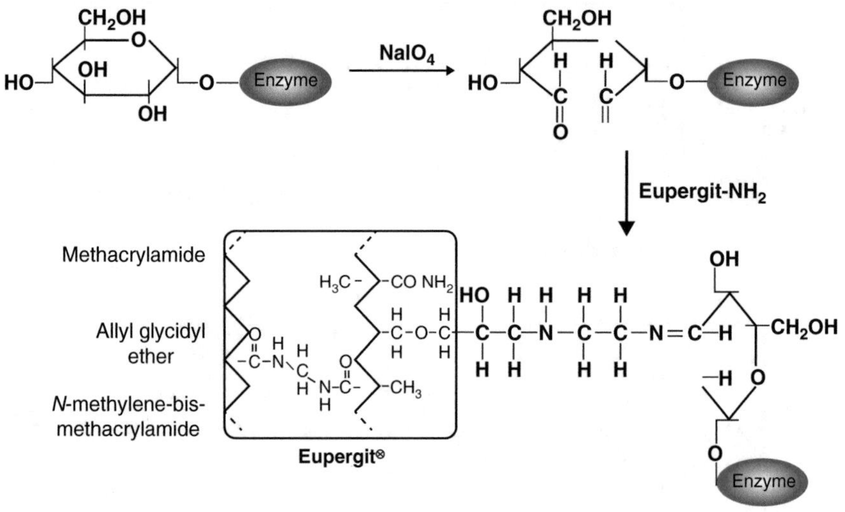

Fig. 3 Schematic illustration of the enzyme oxidation by periodate method and the coupling of the oxidized enzyme to amino-Eupergit support

3.4.1 Oxidation of Enzyme by Sodium Periodate

1. Dissolve 10 mg of lipase in 10 mL of 0.1 M acetate buffer, pH 5.0.
2. Add 500 μL of freshly prepared 0.1 M sodium periodate, and stir the solution (*see* **Note 13**).
3. Protect the solution from light and incubate for 30 min on ice at 4 °C (*see* **Notes 14** and **15**).
4. Stop the reaction by adding 2 mL of 0.01 M ethylene glycol (2 times the amount of NaIO$_4$). Stir the mixture occasionally over a period of 2 h.
5. Remove the by-products by dialyzing the oxidized lipase solution against 50 mM acetate buffer, pH 5.0 (2 L) for 18 h at 4 °C.
6. Measure the volumes before and after dialysis to establish the final concentration of the enzyme.
7. Check the oxidized lipase activity using the method described in Subheading 3.6.

3.4.2 Coupling of Oxidized Enzyme to the Activated Polymers

1. Prepare the reaction mixture containing 1 g of the amino-Eupergit carrier and 20 mL of the dialyzed solution of the oxidized enzyme in sodium acetate buffer at pH 5.0 and stir the mixture for 2 h at 4 °C in the dark (*see* **Notes 16** and **17**).
2. Add 50 μL of freshly prepared sodium borohydride solution in distilled water (4 mg/mL), and stir the mixture occasionally over a period of 4 h at 4 °C (*see* **Note 18**).
3. Remove the by-products and the excess of the sodium *meta*-periodate by dialyzing the solution against 50 mM sodium acetate buffer, pH 5.0 (2 L) overnight at 4 °C.

4. Separate the mixture by vacuum filtration and wash the obtained immobilized enzyme with 100 mL of 0.1 M phosphate buffer, pH 7.0, then three times with 100 mL of the same buffer containing 1 M NaCl.

5. Collect the supernatant and washing buffer solutions for determination of non-immobilized protein and enzyme activity to estimate the efficiency of immobilization (*see* **Note 19**).

3.5 Immobilization via Epoxy Groups with Chemical Amination

The immobilization procedure consists of two main steps: a direct enzyme binding to polymers via their epoxy groups (Subheading 3.1) and the post-immobilization chemical amination of enzyme with ethylenediamine (EDA) after activation of the carboxylic groups with carbodiimide (Fig. 4).

Namely, the use of standard immobilization method is limited because of the problem with the number of enzyme amino groups available for the enzyme attachment. When enzymes are poor in lysine residues, e.g., glycosylated enzymes (Fig. 5), they could be enriched in reactive groups by the chemical amination, increasing the possibilities of having some multipoint covalent attachment [14, 15]. All the surface carboxylic groups of the enzyme seem to be modified with ethylenediamine via carbodiimide chemistry using the advantages of the solid phase chemistry [16, 17].

The amination is carried out as follows:

1. Wash the immobilized lipase onto Eupergit 3 times with an aqueous solution of 1 M EDA (ethylenediamine) and suspended in the same solution (*see* **Notes 20** and **21**). The pH of the solutions is adjusted to 4.75 before suspending the immobilized enzyme and is maintained near this value (±0.1 pH unit) with 0.1 M HCl.

2. Add different amounts of solid EDAC (1-ethyl-3-(dimethylamino-propyl) carbodiimide) to the suspension in portions with constant stirring. Keep the final carbodiimide concentration low (of 10^{-2} or 10^{-3} M) for reasons to be discussed later (*see* **Note 22**).

3. After 90 min of gently stirring at 25 °C, recover the immobilized enzyme beads by filtration on a sintered glass filter by suction under vacuum (*see* **Note 10**).

4. Incubate the recovered immobilized enzyme for 4 h in 0.1 M hydroxylamine solution at pH 7 to recover the modified tyrosines [18].

5. Separate the immobilized enzyme by vacuum filtration and wash the obtained immobilized enzyme three times with 100 mL of the hydroxylamine solution, pH 7.

3.6 Determination of Lipase Activity

Lipase activity can be determined by colorimetric measurement of products of hydrolysis of esters of various fatty acids and

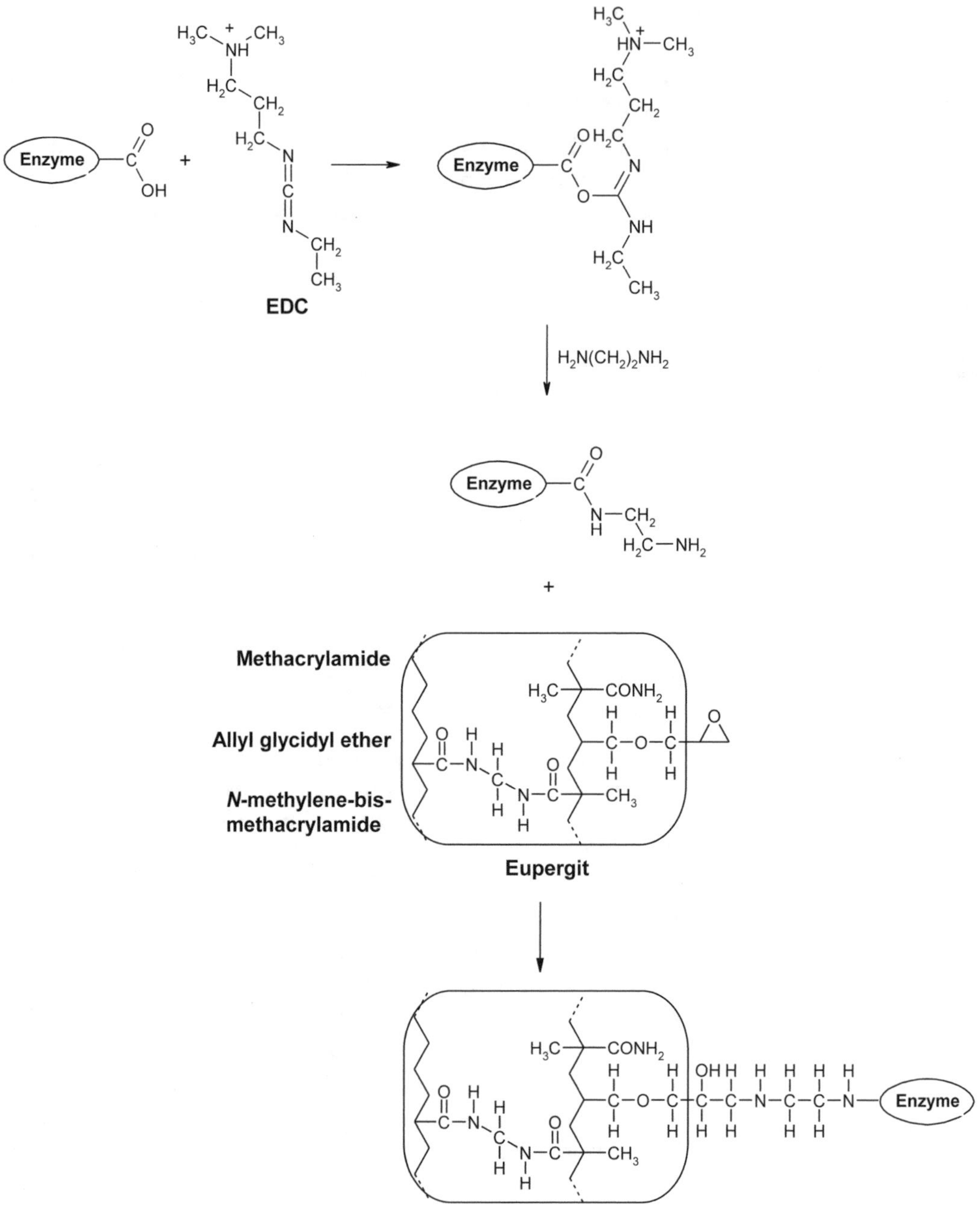

Fig. 4 Schematic illustration of the modified conventional method for enzyme immobilization on epoxy carriers based on the immobilized enzyme post-treatment consisting of the chemical amination of the biocatalyst with ethylenediamine (after activation of the carboxylic function with carbodiimide)

p-nitrophenol, due to the fact that released *p*-nitrophenol strongly absorbs at 410 nm. In this experiment rapid method for lipase activity determination using *p*-nitrophenylpalmitate as substrate was applied [19].

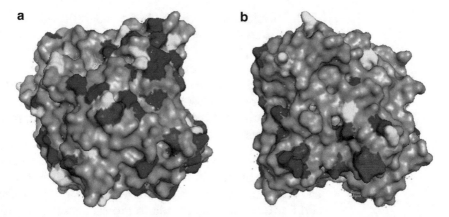

Fig. 5 Distribution of Lys, Asp, and Glu residues on the surface of lipase from *C. rugosa*. Lys is shown in *yellow*; Asp and Glu are shown in *blue* and *red*. (**a**) Front view of catalytic site. (**b**) 90 rotation of the front view in the x-y plane. 3D structure was obtained from Protein Data Bank 1TRH using Pymol

1. Prepare the activity assay mixture: mix 1.8 mL of substrate solution and 50 mg of dry immobilized enzyme, both preincubated at 37 °C.

2. After 1 min of reaction between substrate solution and enzyme at 37 °C, determine the absorbance at 410 nm against blank sample, which contains no enzyme.

3. Prepare the standard graph by using different concentrations of *p*-nitrophenol. Use these data to calculate molar quantity of released reaction product (*p*-nitrophenol). Afterwards, calculate the activity of immobilized enzyme, expressed as IU. One international unit (IU) is defined as the amount of enzyme that liberates 1 μmol of *p*-nitrophenol min^{-1} under assay conditions.

3.7 Determination of Protein Content

Laboratory practice in enzyme immobilization requires a rapid and sensitive method for protein determination to quantify binding capabilities. It is well known that the standard Lowry procedure [20] has disadvantages - it is subject to interference by potassium and magnesium ion, EDTA, Tris, thiol reagents and carbohydrates. Herein the presented Bradford method [21] for protein determination fulfills the requirements for this type of quantification because it eliminates most of the problems with interferences and is easily utilized for processing large numbers of samples as well as adaptable to automation.

The method relies on the change in the absorption spectrum of the dye when bound to the proteins. The protonated form of the dye is a pale orange-red color, whereas the unprotonated form is blue. On binding to proteins in acid solution, the protonation of the dye is suppressed by the positive charges on amino acid side chains (principally arginine) and a blue color results (*see* **Notes 23** and **24**). In addition to these electrostatic effects, hydrophobic interactions between dye and protein are also involved. The protein-dye complex has a high extinction coefficient thus leading to high sensitivity in measurement of the protein. The binding of the dye to protein is a very rapid process (approx. 2 min) and the protein-dye complex remains dispersed in solution for a relatively long time (approx. 1 h, *see* **Note 25**) thus making the procedure very rapid and yet not requiring critical timing for the assay.

1. Standard curve preparation: Pipette BSA solution containing 10–100 µg protein in a volume up to 0.1 mL into test tubes and adjust the volume to 0.1 mL with appropriate buffer. Then add 5 mL of the reagent solution to each test tube and vortex the contents. Allow the mixtures to stand for 5 min at room temperature before the A_{595} is read against a blank containing no protein but 0.1 mL of the appropriate buffer .

2. Protein assay (standard method): Pipette protein solution of unknown sample containing 10–100 µg protein to test tube and adjust the volume to 0.1 mL with buffer. Add 5 mL of the reagent solution to the tube, vortex content, and read A_{595} against a blank after 5 min.

3. Bradford also describes a microprotocol for determination of 1–10 µg protein. Microprotein assay: Pipette protein solution containing 1–10 µg protein in a volume up to 0.1 mL into test tubes and adjust the volume to 0.1 mL with the appropriate buffer. Then add 1 mL of the protein reagent and mix the content. Measure absorbance at 595 nm as in the standard method (but in 1 mL cuvette) against a blank prepared from the protein reagent and 0.1 mL of buffer. Standard curve was prepared and used as in the standard method.

4   Notes

1. All solutions are prepared with distilled and deionized water.

2. All buffer and standard solutions were filtered before use with 0.45 µm Whatman filter paper.

3. Neutral buffers, such as sodium phosphate, pH 7.0, can be used but are not as efficient for oxidation of a glycoenzyme as slightly acidic conditions. Avoid buffers containing primary amines (e.g., Tris) or sugars, which will compete with the intended oxidation and subsequent reactions.

4. Dissolvation of gum arabic in phosphate buffer needs to be performed by slow addition of small portions of gum arabic with stirring. After subsequent addition of Triton X-100, final mixture is mixed in ultrasound water bath for 2 min.

5. In order to obtain translucent substrate solution for lipase activity assay solutions A and B need to be mixed under influence of ultrasound in ultrasound water bath for 2 min.

6. Coomassie brilliant blue R250, used for staining gels, is unsuitable for the preparation of the protein reagent .

7. It is recommended that the same buffer in which the protein is dissolved for the purpose of immobilization is selected. For example, it is 0.1 M phosphate buffer, pH 7.0 used in immobilization procedure II.

8. The immobilization of enzyme on Eupergit via epoxy groups depends on several factors including pH, temperature, ionic strength, coupling time, and enzyme–carrier ratio. Our studies with lipase from *C. rugosa* as a study model show that the amount of the immobilized lipase does not vary significantly over the pH range from 5.0 to 9.5, thus the immobilization reaction is conducted at the optimal value (pH 7.0) for the lipase stability. On the other hand, the immobilization efficiency clearly depends on the molarity (ionic strength) of the buffer used for the immobilization procedure, the coupling time and the enzyme–carrier ratio. At optimal pH and the buffer concentration (pH 7.0, 1.0 M), the maximal immobilized lipase activity was obtained after 48 h coupling at 25 °C at the enzyme–carrier ratio of 1:10. It appears that it is necessary to use a high ionic strength to promote the covalent immobilization of enzymes to Eupergit® supports by conventional method. Some enzymes cannot be exposed to these conditions, and not all enzymes seem to be immobilized even under those conditions.

9. An alternative procedure for immobilization via epoxy groups (Subheading 3.1) was developed recently with the intention to reduce immobilization time [22]. This procedure differs from the procedure described in Subheading 3.1 only in **step 1**. Instead of 48 h contact between enzyme and support at 4 °C, an alternative method of immobilization is carried out under the influence of microwave irradiation. Put mixture of Eupergit® and 1.0 M sodium phosphate buffer (same amounts as in Subheading 3.1) in round bottom flask (volume 100 mL) and place flask in microwave oven and adjust power of irradiation at 100 W and irradiate mixture for 2 min. It must be emphasized that optimal immobilization time (2 min) was determined with immobilized lipase activity as criterion, and in case of another enzyme the immobilization time should be optimized before use. Also, pay attention to the fact that action of microwave irradiation is strongly influenced by shape of flask and volume of applied liquid. Any variation will lead to different immobilization results .

10. The pore size of the sintered glass filter is recommended to be 40–100 µm since the size of Euppergit particles is around 150 µm.

11. Modification degree could be quantified by titration of the amino groups introduced in the support [23].

12. Aldehyde derivatives react with amino groups on the enzyme to form a Schiff's base linkage. Sulfhydryl and imidazole groups may undergo similar reactions. One disadvantage of this method is the reversibility of Schiff's base formation in aqueous media, especially at low pH values. This can be overcome by reduction of the imine bonds with sodium borohydride to yield stable secondary amine linkage.

13. The immobilization of enzyme on Eupergit by periodate protocol can be adapted for many other glycoenzymes. Excessive oxidation of enzyme can lead to loss of enzyme activity, thus the reaction conditions are critical. The degree of oxidation could be controlled by the amount of periodate used and the time allowed for the reaction. For lipase from *C. rugosa*, the oxidant concentration of 5 µmol per mg enzyme preparation and a 30-min reaction period seem to be the most suitable. At these conditions, the specific activity of the oxidized lipase is almost the same as the original [10].

14. Preparations of lipases may vary in their carbohydrate content, and this can affect the degree of oxidation. Free carbohydrate can be removed by gel filtration. Increasing the sodium periodate amount to 10 µmol/mg can be beneficial, but further increases lead to the enzyme inactivation.

15. The reaction is light sensitive and should be performed in the dark.

16. The amount of protein bound to the amino-Eupergit supports will depend on the coupling reaction conditions used, e.g., time, temperature, pH, the degree of enzyme oxidation and its amount in the coupling solution. While the optimal coupling conditions have been determined for common enzymes such as lipases, these same conditions may not hold true for other enzymes, and optimized studies may be necessary.

17. Studies have been carried out with common enzymes to determine the binding capacity of Eupergit® supports including lipases [10], α- and β-galactosidase [7], pepsin, trypsin [6], penicillin G acylase [8], and others. In general, Eupergit supports bind from 2 to 90 mg protein per gram of support. However, too high binding capacities could result in steric interference problems, loss of enzyme activity, and diffusion limitation. Diffusion problems become more significant with increasing the enzyme amount on the beads. A suitable lipase

amount is around 15.6 mg/g resulting in a rather high activity yield of 79 % with the satisfactory degree of enzyme fixed [10].

18. Amines can react with aldehyde groups under reductive amination conditions using a suitable reducing agent such as sodium borohydride. The result of this reaction is a stable secondary amine linkage. Alternatively, amines spontaneously react with aldehydes to form covalent linkages, although the addition of a reducing agent increases the efficiency of the reaction. Warning: sodium cyanoborohydride is toxic, prepare solutions and perform reactions in a fume hood.

19. For short-term storage, the enzyme-immobilized beads are kept in 0.1 M potassium phosphate, pH 7.0, at 4 °C. For long-term storage, the beads can be quickly washed with water and air-dried (this procedure is not tolerated by very sensitive enzymes).

20. Carbodiimide-promoted activation of carboxylic groups in enzymes can lead to quantitative or selective modification through condensation with a nucleophile (in our case an amine) (see Fig. 4). In the absence of the exogenous nucleophile, the formation of linkages between enzyme carboxylic and amino groups can be formed, leading to intramolecular cross-linking. This problem appears to be overcome by using a water-soluble carbodiimide to activate the carboxyl functions in the enzyme at pH 4–5 and a suitable buffer . Avoid buffers containing amines such as Tris or glycine; acetate buffers are also not recommended. Phosphate buffers reduce coupling efficiency, which can be compensated by increasing the concentration of carbodiimides. MES buffer (4-morpholinoethanesulfonic acid) is a suitable carbodiimide reaction buffer.

21. A large excess of ethylenediamine is employed in the first step in order to modify all the surface carboxylic groups of the enzyme. The highest increase of enzyme activity and stability is obtained by the full modification of the enzyme surface.

22. Careful control of the reaction conditions and choice of carbodiimide allow a great degree of selectivity in this reaction. The activation of the enzyme and support should be optimized, with respect to activity yield and stability of the immobilized enzyme. Low amount of added EDAC can lead to insufficient immobilization of enzyme, while conversely high concentration can lead to unwanted cross-linking of enzyme and alteration of catalytic conformation [15]. The pH is also a critical factor for this step. At pH ranging from 4 to 5, the reactivity of the proximal (reactive) amino group is considerably enhanced with respect to the amino group remaining after attachment.

23. Some variation in response between different proteins is present which can be correlated to some extent with the content of basic side chains.

24. The presence of Triton and SDS at the concentration 1 % (but not at the 0.1 %) leads to interferences and error in measurements.

25. The dye–protein complex tends to precipitate over periods longer than an hour. This is more of a problem with quartz cuvettes, so glass or plastic ones should be used.

Acknowledgments

This work was supported by the Ministry of Science and Technological Development of the Republic of Serbia (Project No III46010).

References

1. Sheldon RA (2007) Enzyme immobilization: the quest for optimum performance. Adv Synth Catal 349:1289–1307. doi:10.1007/s10562-014-1406-2

2. Cantone S, Ferrario V, Corici L et al (2013) Efficient immobilisation of industrial biocatalysts: criteria and constraints for the selection of organic polymeric carriers and immobilisation methods. Chem Soc Rev 42:6262–6276. doi:10.1039/c3cs35464d

3. Balcao VM, Paiva AL, Malcata XF (1996) Bioreactors with immobilized lipases: state of the art. Enzyme Microb Technol 18:392–416. doi:10.1016/0141-0229(95)00125-5

4. Saponjić S, Knežević-Jugović ZD, Bezbradica DI et al (2010) Use of Candida rugosa lipase immobilized on sepabeads for the amyl caprylate synthesis: batch and fluidized bed reactor study. Electron J Biotechn 13. DOI: 10.2225/vol13-issue6-fulltext-8

5. Warmerdam A, Benjamins E, De Leeuw TF et al (2014) Galacto-oligosaccharide production with immobilized galactosidase in a packed-bed reactor vs. free -galactosidase in a batch reactor. Food Bioprod Process 92:383–392. doi:10.1016/j.fbp.2013.08.014

6. Katchalski-Katzir E, Kraemer DM (2000) Eupergit® C, a carrier for immobilization of enzymes of industrial potential. J Mol Catal B Enzym 10:157–176. doi:10.1016/S1381-1177(00)00124-7

7. Hernaiz MJ, Crout DHG (2000) Immobilization/stabilization on Eupergit C of the β-galactosidase from B. circulans and an α-galactosidase from Aspergillus oryzae. Enzyme Microb Technol 27:26–32. doi:10.1016/S0141-0229(00)00150-2

8. Rocchietti S, Urrutia SVA, Pregnolato M et al (2002) Influence of the enzyme derivative preparation and substrate structure on the enantioselectivity of penicillin G acylase. Enzyme Microb Technol 31:88–93. doi:10.1016/S0141-0229(02)00070-4

9. Tibhe JD, Fu H, Noël T et al (2013) Flow synthesis of phenylserine using threonine aldolase immobilized on Eupergit support. Beilstein J Org Chem 9:2168–2179. doi:10.3762/bjoc.9.254

10. Knežević Z, Milosavić N, Bezbradica D et al (2006) Immobilization of lipase from Candida rugosa on Eupergit® C supports by covalent attachment. Biochem Eng J 30:269–278. doi:10.1016/j.bej.2006.05.009

11. Hilal N, Kochkodan V, Nigmatullin R et al (2006) Lipase-immobilized biocatalytic membranes for enzymatic esterification: comparison of various approaches to membrane preparation. J Membr Sci 268:198–207. doi:10.1016/j.memsci.2005.06.039

12. Mateo C, Abian O, Bernedo M et al (2005) Some special features of glyoxyl supports to immobilize proteins. Enzyme Microb Technol 37:456–462. doi:10.1016/j.enzmictec.2005.03.020

13. Žuža M, Milosavić N, Knežević-Jugović Z (2009) Immobilization of modified penicillin G acylase on Sepabeads carriers. Chem Pap 63:117–124. doi:10.2478/s11696-009-0012-z

14. Rueda N, Santos JCSD, Ortiz C et al (2015) Chemical amination of lipases improves their immobilization on octyl-glyoxyl agarose beads. Catal Today. doi:10.1016/j.cattod.2015.05.027

15. Bezbradica D, Jugović B, Gvozdenović M et al (2011) Electrochemically synthesized polyaniline as support for lipase immobilization. J Mol Catal B Enzyme 70:55–60. doi:10.1016/j.molcatb.2011.02.004

16. Oliveira GB, Carvalho LB, Silva MPC (2003) Properties of carbodiimide treated heparin. Biomaterials 24:4777–4783. doi:10.1016/S0142-9612(03)00376-4

17. Rodrigues RC, Berenguer-Murcia Á, Fernandez-Lafuente R (2011) Coupling chemical modification and immobilization to improve the catalytic performance of enzymes. Adv Synth Catal 353:2216–2238. doi:10.1002/adsc.201100163

18. Carraway KL, Koshland DE Jr (1968) Reaction of tyrosine residues in proteins with carbodiimide reagents. Biochim Biophys Acta 160:272–274

19. Licia MP, Cintia MR, Mario DB et al (2006) Catalytic properties of lipase extracts from *Aspergillus niger*. Food Technol Biotechnol 44:247–252

20. Lowry OH, Rosebrough NJ, Farr AL et al (1951) Protein measurement with Folin-phanol reagent. J Biol Chem 193:265–275

21. Bradford MM (1976) A rapid and sensitive method for the quantification of microgram quantities of protein utilizing the principle of protein-dye binding. Anal Biochem 72:248–254

22. Bezbradica D, Mijin D, Mihailović M, Knežević-Jugović Z (2009) Microwave-assisted lipase immobilization of lipase from *Candida rugosa* on Eupergit® supports. J Chem Technol Biotechnol 84:1642–1648. doi:10.1002/jctb.2222

23. Fernandez-Lafuente R, Rosell CM, Rodríguez V et al (1993) Preparation of activated supports containing low pK amino groups. A new tool for protein immobilization via the carboxyl coupling method. Enzyme Microb Technol 15:546–550. doi:10.1016/0141-0229(93)90016-U

Chapter 8

Micellar Polymer Encapsulation of Enzymes

Sabina Besic and Shelley D. Minteer

Abstract

Although enzymes are highly efficient and selective catalysts, there have been problems incorporating them into fuel cells. Early enzyme-based fuel cells contained enzymes in solution rather than immobilized on the electrode surface. One problem utilizing an enzyme in solution is an issue of transport associated with long diffusion lengths between the site of bioelectrocatalysis and the electrode. This issue drastically decreases the theoretical overall power output due to the poor electron conductivity. On the other hand, enzymes immobilized at the electrode surface have eliminated the issue of poor electron conduction due to close proximity of electron transfer between electrode and the biocatalyst. Another problem is inefficient and short term stability of catalytic activity within the enzyme that is suspended in free flowing solution. Enzymes in solutions are only stable for hours to days, whereas immobilized enzymes can be stable for weeks to months and now even years. Over the last decade, there has been substantial research on immobilizing enzymes at electrode surfaces for biofuel cell and sensor applications. The most commonly used techniques are sandwich or wired. Sandwich techniques are powerful and successful for enzyme immobilization; however, the enzymes optimal activity is not retained due to the physical distress applied by the polymer limiting its applications as well as the non-uniform distribution of the enzyme and the diffusion of analyte through the polymer is slowed significantly. Wired techniques have shown to extend the lifetime of an enzyme at the electrode surface; however, this technique is very hard to master due to specific covalent bonding of enzyme and polymer which changes the three-dimensional configuration of enzyme and with that decreases the optimal catalytic activity. This chapter details encapsulation techniques where an enzyme will be immobilized within the pores/pockets of the hydrophobically modified micellar polymers such as Nafion® and chitosan. This strategy has been shown to safely immobilize enzymes at electrode surfaces with storage and continuous operation lifetime of more than 2 years.

Key words Nafion®, Chitosan, Micellar polymers, Enzyme immobilization, Enzyme encapsulation, Enzyme stabilization, Biosensors, Biofuel cells

1 Introduction

One of the primary causes of protein instability is their susceptibility to conformational change around the active site and subsequent inactivation or denaturation. The structure of a protein is arranged by the intramolecular interactions and its interactions with its surroundings such as ionic bonds, hydrogen bonds, covalent bonds, polar and Van der Waals forces, hydrophobic and electrostatic

Shelley D. Minteer (ed.), *Enzyme Stabilization and Immobilization: Methods and Protocols*, Methods in Molecular Biology, vol. 1504, DOI 10.1007/978-1-4939-6499-4_8, © Springer Science+Business Media New York 2017

interactions. Primary structure of a protein is an amino acid sequence that is linked with strong peptide bonds that are usually unbreakable during hydration or temperature changes. Secondary structure consists of several repeating patterns, such as α-helix and the β-pleated sheet, stabilized by hydrogen bonds between the carbonyl and N-H groups in the polypeptide's backbone. During the preservation process or storage, the secondary structure pattern ratios are altered; however, ultimately, it is the tertiary structure of the protein that determines its function. The tertiary structure is a three-dimensional conformation where interactions between the side chains in the primary structure take place. Interactions such as: hydrophobic and electrostatic interactions as well as hydrogen and covalent bonding. Also, the hydrophobic side groups are being hidden in the core and the hydrophilic groups are exposed to the surrounding environment (*see* Fig. 1). Therefore, proteins are easily being denatured by strong acids or bases altering hydrogen bonding and salt bridge patterns, organic solvents such as ethanol and detergents interfere with hydrophobic interactions, reducing agents such as urea and β-mercaptoethanol break disulfide bridges converting them into sulfhydryl groups, high salt concentration will aggregate the protein and precipitate it out of the solution, heavy metal ions form ionic bonds with negatively charged groups resulting in a significant conformational tertiary alteration, temperature changes change the molecular vibration eventually weakening the hydrogen bonds and mechanical stress disrupts the delicate balance of forces that maintain protein structure. Most

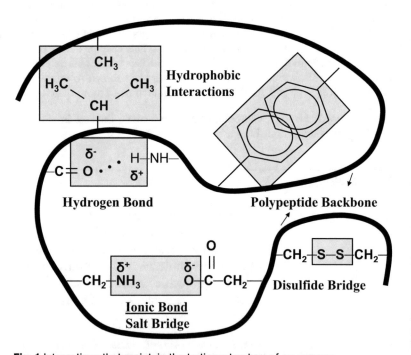

Fig. 1 Interactions that maintain the tertiary structure of an enzyme

commonly utilized protein stabilization methods employed today are based on dampening the molecular motions and therefore eliminating conformational transitions while the protein is still in the native state [1, 2].

Although enzymes are highly efficient and selective catalysts, there have been problems incorporating them into fuel cells. Early enzyme-based fuel cells contained enzymes in solution rather than immobilized on the electrode surface (*see* Fig. 2) [3]. One problem utilizing an enzyme in solution besides the issue of maintaining the catalytic activity is an issue of transport associated with long diffusion lengths between the site of bioelectrocatalysis and the electrode. This issue drastically decreases the theoretical overall power output due to the poor electron conductivity. On the other hand, enzymes immobilized at the electrode surface have eliminated the issue of poor electron conduction due to close proximity of electron transfer between electrode and the biocatalyst. The major problem is inefficient and short term stability of catalytic activity within the enzyme that is suspended in free flowing solution. Enzymes in solutions are only stable for hours to days, whereas immobilized enzymes can be stable for weeks to months and now even years. Over the last decade, there has been substantial research on immobilizing enzymes at electrode surfaces for biofuel cell and sensor applications [4–8]. The most commonly used techniques are sandwich [9] or wired [7, 10, 11] (*see* Figs. 3 and 4). Sandwich techniques are powerful and successful for enzyme immobilization; however, the enzymes optimal activity is not retained due to the physical distress applied by the polymer limiting its applications as well as the non-uniform distribution of the enzyme and the

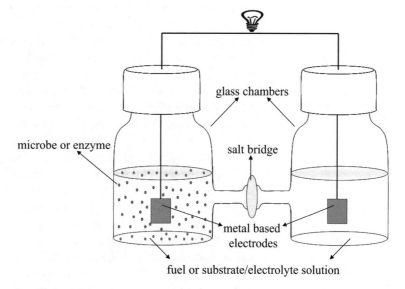

Fig. 2 Example biofuel cell containing microbes or enzymes suspended in solution

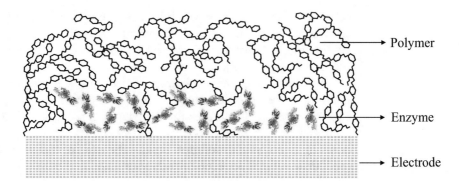

Fig. 3 Sandwich technique. Enzymes are immobilized between the electrode and the polymer

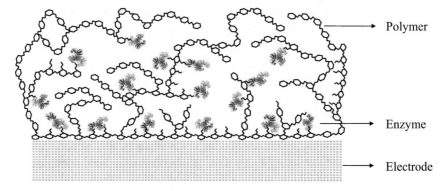

Fig. 4 Wired technique. Enzymes are immobilized via covalent bond to the polymer

diffusion of analyte through the polymer is slowed significantly [5, 10–12]. Wired techniques have been shown to extend the lifetime of an enzyme at the electrode surface; however, this technique is very hard to master due to specific covalent bonding of enzyme and polymer which changes the three-dimensional configuration of enzyme and with that decreases the optimal catalytic activity [7]. In this chapter, we detail an encapsulation technique (*see* Fig. 5) where an enzyme is immobilized within the pores of the hydrophobically modified micellar polymers such as Nafion® and chitosan. Hydrophobic micelles will minimize the motion of an entire enzyme unit forcing it to stay and maintain its native state via repelling forces between the native hydrophilic exterior of an enzyme and the hydrophobic side chains. Further benefits of encapsulating enzymes within the micellar polymers are the minimization of the high concentration of the denaturing factors such as drastic changes in pH, organic solvents, detergents, reducing agents, salt concentrations, and heavy metals increasing the lifetime of the catalytic activity within the enzyme. Additionally, the micellar polymer further provides the insulation from drastic temperature changes as well as protection against various mechanical stresses minimizing the denaturing process. This strategy has shown to safely immobilize

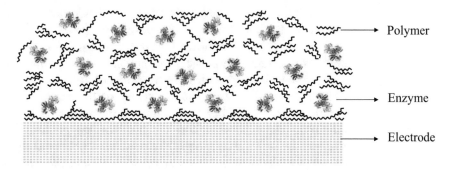

Fig. 5 Encapsulation technique. Enzymes are immobilized within the micellar pores/pockets of the micellar polymer

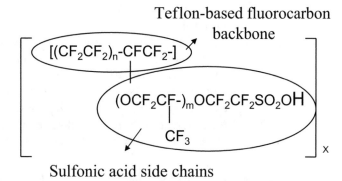

Fig. 6 Nafion® 117, where $m=1$ $n=6, 7, \ldots, 14$. Molecular structure of a commercial unmodified micellar polymer. Acidic proton (the *bolded* H) is where the hydrophobic salts exchange with the proton in the process of modification

enzymes at electrode surfaces with storage and continuous operation lifetime of more than 2 years [8].

Polymers with distinct hydrophobic and hydrophilic domains can naturally form micellar structures that are more stable than surfactant micelles [13, 14]; therefore, hydrophobically modified micellar polymers will form secure and durable pocket structures for long-term enzyme stabilization, immobilization, and protection against harsh chemical and physical forces of the surrounding environment extending the lifetime of their catalytic activity.

Nafion® is a perfluorinated ion exchange polymer with a micellar pore/pocket structure that has excellent properties as an ion conductor and has been widely employed to modify electrodes for variety of sensor and fuel cell applications. The molecular structure of commercial Nafion® is shown in Fig. 6. More specifically, Nafion® is a cation-exchange polymer that has superselectivity against anions and the ability to preconcentrate cations at the electrode surface which in turn give rise to cation selective systems. This property of superselectivity against anions comes from high ionic charge density at the orifices of the small channels [15]. Even though Nafion® is a micellar polymer, it has not been successful at immobilizing biological entities at the electrode surface due to the extremely small and super acidic pore structures. Previous studies by Minteer et al. have shown that mixture-cast films of quaternary ammonium salts (such as: tetrabutylammonium bromide) and Nafion® have increased the mass transport of small analytes through the films and decreased the selectivity of the membrane against anions [16]. Quaternary ammonium bromide salts have higher affinity to the sulfonic acid side chain than the proton; therefore, they can be utilized to modify the polymer and extend the lifetime of the modified Nafion® because protons are less likely to exchange back into the membrane and re-acidify it. Therefore, the modified Nafion® will have a higher buffering capacity when subjected to extreme acidic and basic solutions providing an exceptional physiological environment for enzyme immobilization. Another advantage of the "high affinity property" of sulfonic acid group to large hydrophobic species is a simple technique by which the pores of the polymer are easily altered into different sizes (*see* Fig. 7) [17] and shapes further enhancing the ability of the enzyme entrapment as well as retaining its catalytic activity. Nafion® modified with quaternary ammonium bromide salts will not only provide a buffered micellar structure for easier enzyme immobilization [18, 19], but

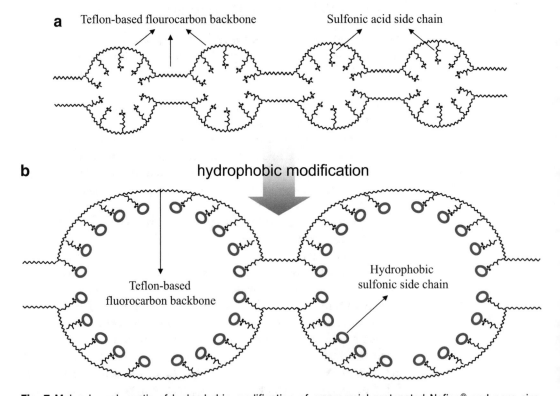

Fig. 7 Molecular schematic of hydrophobic modification of commercial protonated Nafion® and pore size alteration (not to scale). (**a**) Commercial (protonated) unmodified micellar polymer, Nafion®. (**b**) Hydrophobically modified micellar polymer, Nafion®

will also retain the electrical properties of unmodified Nafion® as well as increase the mass transport flux of ions and neutral species through the membrane minimizing problems associated with slow diffusion [14–16, 20–22]. Even though Nafion® is widely used and studied by various scientists and manufacturers due to the fact that it possess excellent properties that are easily altered into new desired properties with high specificity and efficiency for various chemical and biological applications, it is still considered as an expensive perfluorinated ionomer which is neither fully biocompatible nor biodegradable.

On the other hand, chitosan is a typical marine linear polysaccharide formed from chitin and next to cellulose, it is the second most abundant biomass on earth found in the shells of crustaceans and insects [23, 24]. The molecular structure of chitosan is shown in Fig. 8. Because cellulose and chitin are produced in the largest amounts, they are considered as the most abundant organic compounds on earth estimated to be around 10^{11} tons each year [25]. Chitosan exhibits polyelectrolyte behavior due to the protonated amino groups in the polymeric backbone [26]. It is also well known for its biocompatibility, biodegradability and nontoxic properties as well as for chemical inertness and high mechanical strength. Due to excellent properties of chitosan, it has been adopted into various distinctive biological and physicochemical applications such as adsorption of metal ions as cleansing mechanism for waste water, coagulation of suspensions or solutes, film- and fiber forming properties, as well as antibacterial, antifungal, healing, and drug delivery for pharmaceutical industry [25, 27, 28]. Chitosan also has phytosanitary effects on plants and humans where it appears to help reduce blood cholesterol levels [27] as well as being water insoluble [29]. Because of all of these properties, chitosan grew interests and became an attractive polymer for various sensing applications where electrodes were modified with chitosan to preconcentrate both cations and anions or to immobilize enzymes and proteins for biosensing [30–36]. Chitosan is easily hydrophobically modified by reductive amination using aldehydes with long alkyl chain lengths [29] and similar to Nafion® it forms polymer micelles

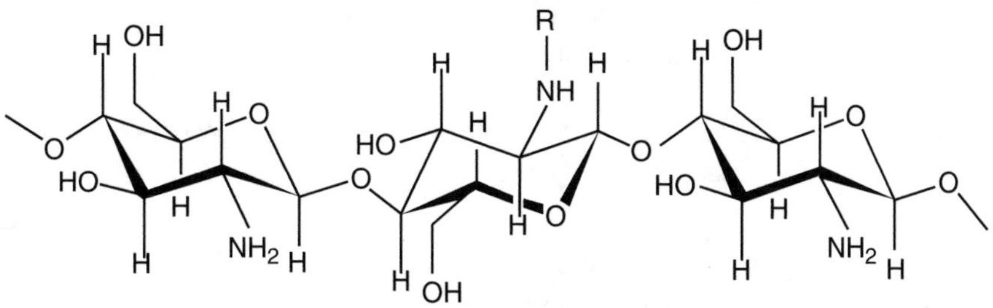

Fig. 8 Chitosan. Molecular structure of unmodified chitosan; where R = butyl, hexyl, octyl, or decyl alkyl chain after modification

due to the separate hydrophobic and hydrophilic segments of the polymer [26, 37, 38]. Therefore, hydrophobically modified chitosan have advantageous potential for renewable energy, such as: biofuel cell applications [39].

2 Materials

2.1 TMICA (Trimethylicosyl-ammonium Bromide); Synthesized

1. 1-bromoeicosane: 98 % pure.
2. Trimethylamine anhydrous: 99 % pure.
3. Teflon bomb (276ACT304 012506 manufactured by Parr Instrument Company, Moline, IL).

2.2 Modification of Nafion®

1. Nafion®: perfluorinated resin solution 5 % by weight in lower aliphatic alcohols and water, contains 15–20 % water (Aldrich).
2. Ethanol: 190 proof, 95.0 % pure.
3. Propanol: 99.5 % pure.
4. Quaternary ammonium salts.
   (a) T3A (tetrapropylammonium bromide): 98 % pure.
   (b) TBAB (tetrabutylammonium bromide): 99 % pure.
   (c) T5A (tetrapentylammonium bromide): 99 % pure.
   (d) TEHA (triethylhexylammonium bromide): 99 % pure.
   (e) TMHA (trimethylhexylammonium bromide): 98 % pure.
   (f) TMOA (trimethyloctylammonium bromide): 98 % pure.
   (g) TMDA (trimethyldecylammonium bromide): 98 % pure.
   (h) TMDDA (trimethyldodecylammonium bromide): 99 % pure.
   (i) TMODA (trimethyloctadecylammonium bromide): 98 % pure.
   (j) TMTDA (trimethyltetradecylammonium bromide): 99 % pure.
   (k) TMHDA (trimethylhexadecylammonium bromide): 98 % pure in powder form.
   (l) TMICA (trimethylicosylammonium bromide): synthesized in the lab.
5. Water: 18 MΩ used for rinsing and extracting the excess salt from the polymer.

2.3 Modification of Chitosan

1. Chitosan- low molecular weight (LMW) and medium molecular weight (MMW): from crab shells, 75–85 % deacetylated (Aldrich).
2. Acetic acid: 99.7 % pure liquid.
3. Methanol: 99.8 % pure.
4. Butanal: 98 % pure.

5. Hexanal: 98 % pure.

6. Octanal: 99 % pure liquid.

7. Decanal: 95 % pure liquid.

8. Sodium cyanoborohydride: 95 % pure.

9. Tert-amyl alcohol: 99 % pure liquid.

10. Chloroform: 99 % pure preserved with 0.75 % ethanol.

2.4 Encapsulation of Enzymes Within the Pores of Modified Nafion® and/or Chitosan

1. Modified Nafion® and/or chitosan: Keep tightly closed and away from heat, sparks, and open flame.

2. Enzyme in powder or solution form: store as required.

3 Methods

3.1 Synthesis of TMICA (Tri-methylicosylammonium Bromide)

1. Add 2.5 g (0.006917 mol) of 1-bromoeicosane into the 23 mL Teflon cup and cap it closed.

2. Place the closed Teflon cup into a Teflon bomb container and chill it cold for about 5 min with dry ice.

3. Pour 5 mL of chilled trimethylamine (in molar excess to the alkylbromide) into the chilled Teflon bomb container containing chilled 1-bromoeicosane.

4. Seal the Teflon bomb with screw cap and place in the oven at 100 °C for 90 min.

5. Move it to the hood and cool down to room temperature. With a fan, it takes about an hour. After cooling down, slowly open until the air within the container is depressurized, or after the sound of air suction.

6. Transfer the product into the dark glass container and place it into the vacuum desiccator for 2 h to remove any excess tri-methylamine that might be dissolved in the product.

7. Tightly close the container with a lid and store it in a dry and well ventilated place.

8. The product is in the form of a white powder.

3.2 Preparation of Modified Nafion® with Quaternary Ammonium Salts

Enzyme encapsulation within the micellar polymer is obtained via modified Nafion®. Treating the polymer by various quaternary salts alters the pores to different shapes and sizes corresponding to specific structure of each salt and providing the optimal catalytic environment for various enzymes. The amount of salt used in all mixture-casting solutions is prepared such that the concentration of quaternary ammonium salt is in threefold excess of the concentration of sulfonic acid sites in the Nafion® suspension. After the modified polymer has been obtained, then the encapsulation of an enzyme within the polymer takes place. Bilirubin oxidase and glyceraldehyde-3-phosphate dehydrogenase are very small enzymes (~50 kDa) and shown to best encapsulated via TBAB modified Nafion. Alcohol dehydrogenase of similar size and shape have also shown to be best immobilized within the pores of TBAB modified Nafion. Larger size and shape of catalyst prefer larger pore sizes for retention of their optimal catalytic activity. Refer to Table 1 for additional examples of optimal enzymes stability within a specific polymer.

1. Add 4.052×10^{-4} mol of salt per 1 mL of 5 % by weight Nafion® suspension and mix by vortex for ~10 min, or until homogeneous mixture is obtained and salt has dissolved completely.

2. Co-cast the quaternary ammonium salt with Nafion® suspension into a weighing boat. It is very important not to cast more than 1 mL per 8 cm² surface area (*see* **Note 1**).

3. Let the mixture air dry in low humidity for 12–24–48 h (*see* **Note 2**).

4. Once dry, soak the mixture-cast film in 18 MΩ water for at least 24 h to remove all excess salts (*see* **Note 3**).

5. After the salt extraction, rinse the films thoroughly with 18 MΩ water three times (*see* **Note 4**).

6. After thoroughly rinsing and salt extraction, allow the modified polymer to air dry in two half closed containers to avoid any particle or dust contamination.

7. Qualitatively transfer the dry films into a glass vial and resuspend the films into 1 mL of lower aliphatic alcohols to prepare the suspension for enzyme immobilization (*see* **Note 5**).

8. Vortex the modified polymer suspension for about ~10 min and store at 25 °C tightly closed with a cap and additionally sealed with Parafilm (*see* **Note 6**).

3.3 Preparation of Hydrophobically Modified Chitosan via Reductive Aldehyde Amination

Before enzyme encapsulation takes place, the chitosan, a linear polysaccharide, needs to be transformed into a micellar polymer via reductive amination. In this reaction the amine group on the chitosan will attack the carbon on the aldehyde forming an intermediate with a double covalent bond between each other and water as a by-product. Utilizing different sizes of aldehydes such as butanal, octanal or decanal, the micellar shape, size, and distribution is altered. After the modified polymer has been obtained, then the encapsulation of an enzyme within the polymer takes place. Refer to Table 1 for additional examples of optimal enzymes stability within a specific polymer.

Table 1
Optimum polymers for immobilization of several enzymes

Enzyme	Optimum modified polymer
Alcohol dehydrogenase Bilirubin oxidase Glyceraldehyde-3-phosphate dehydrogenase	TBAB* modified Nafion®
Lipoxygenase	TMDDA* modified Nafion®
Hexokinase, phosphofructose kinase	TMHDA* modified Nafion®
Phosphoglucose isomerase	TMHA, TMOA, T3A, and TBAB modified Nafion®
Aldolase	TMTDA* modified Nafion®
Glucose dehydrogenase	Butyl* modified chitosan TMOA modified Nafion®
Formate dehydrogenase	TMHA* modified Nafion® butyl modified chitosan
Aldehyde dehydrogenase	Hexyl and butyl* modified chitosan TMHA and TMOA modified Nafion®
Phosphoglycerate kinase	All salts ref. for modification of Nafion® except TMDDA and TMTDA
Enolase	Hexyl and butyl modified chitosan
Pyruvate kinase	TMODA* modified Nafion®
PQQ-dependent glucose dehydrogenase	TMODA* modified Nafion®
PQQ-dependent alcohol dehydrogenase	TMOA*, TEHA, TBAB modified Nafion®

The symbol *asterisk* (*) labels the polymer that have shown the best ability to immobilize the enzyme as compared to other polymers studied here

1. Dissolve medium molecular weight chitosan (0.500 g) in 15 mL of 1% (v/v) acetic acid by rapid stirring for approximately 10 min, forming a viscous gel-like solution.

2. After the formation of a homogeneous gel-like solution, add 15 mL of absolute methanol to the chitosan mix and stir for approximately 15 min (*see* **Note 7**).

3. To hydrophobically modify chitosan, slowly add 15 mL of aldehyde of choice (butanal, hexanal, octanal, or decanal) over a period of 1 min while continue on stirring (*see* **Note 8**).

4. Also add 1.25 g of sodium cyanoborohydride while the gel is continuously stirred and stir until the suspension is cooled to room temperature (*see* **Note 9**).

5. Collect the resulting product in the form of a white precipitate by vacuum filtration through a Buchner funnel (*see* **Note 10**).

6. Wash with 150 mL increments of absolute methanol three to four times to remove any unreacted cyanoborohydride.

7. Dry the modified chitosan over a filter and then transfer to a watch dish to be further dried in a vacuum desiccator over at 40 °C for 2 h, leaving a flakey white solid (*see* **Note 11**).

8. Store the modified chitosan in the form of a flakey white solid tightly closed in a glass vial for up to 6 months.

9. Resuspend in either ethanol or n-propanol (*see* **Note 12**).

3.4 Encapsulation of Enzymes Within the Pores of Modified Nafion® and or Chitosan

There are two enzyme encapsulation methods. One of them is where the enzyme in powder form is directly added to the modified Nafion® while the second method is where enzyme in buffer is added to the modified Nafion® suspension. Usually 1 mg of enzyme is used regardless of method; however, larger or smaller amounts have been used and shown proper enzyme immobilization. Depending on the enzymes stability within an alcohol environment is what determines the ratio of enzyme solution to the modified polymer solution. Ratios such as 1/2, 2/3, and even 1/3 of polymer have been tested and shown that in all cases the polymer has been able to properly dry and form micelles that efficiently immobilize the enzymes at the electrode surface.

1. Mix the enzyme with the modified polymer.

2. Vortex for ~20 min, or until the homogeneous mixture has been achieved to ensure the complete encapsulation of all enzymes introduced to the polymer (*see* **Note 13**).

3. Cast the enzyme/membrane suspension onto a conductive surface and let air dry over night for 12–24–48 h at low humidity (*see* **Note 14**).

4. Once dry, an enzyme encapsulation within the pores of the polymer is complete and ready for use. They can be stored as is or used immediately. Rehydration of the polymer must be done at least 1 h before any experiments are performed; however, overnight soak seems to be sufficient enough for more reproducible results. Rehydration can be done with either only buffer or fuel/substrate solution.

4 Notes

1. To thick or thin films do not dry properly and therefore do not form uniform micelles within the polymer.

2. Do not dry it over a running fan or any noticeable air movement. Micelles do not form properly if the casting is dried too fast. Therefore, half close the casted polymer weighing boat with another empty weighing boat and place it into a half closed box to minimize the air and dust movement that might alter and disrupt proper micelle formation. Keep the whole

setup above or close to dehumidifier to better control low humidity that will help in consistent micelle formation (Whirlpool dehumidifier set on continuous-high level). Depending on humidity of the environment, it can still take between 12 and 48 h to completely dry, but if humidity is above average high it will never dry. During winter time, it usually takes at least 12 h to dry due to very low humidity.

3. The dry film will have some white and yellow spots or discoloration indicating where excess salt is situated. After all salt is removed, the film becomes more transparent; however, it is very hard to get a crystal clear film which would be a perfect micelle formation. During these first 24 h, if the perfect film has formed then the film itself will rise and float. Cloudy or white membranes either did not dry properly or the micellar structure has been destroyed. If the mixture-cast film does not float, try to peel it of the surface. Do not scrape the surface to obtain your polymer. If necessary cast thicker films so they do not break during peeling process, but make sure they are not too thick because of the improper micelle formation. To start the peeling process, make sure the polymer is hydrated with 18 MΩ H_2O and fold over the weighting boat couple of times destructing the interactions mechanically between the polymer and the weighting boat. This will release the polymer and make the peeling process much easier. This can only be done in plastic weight boats, and not glass. For the polymers that did not float during the first 24 h and needed to be peeled off, they need further salt extraction for 24 h before resuspension into solvents.

4. Pay careful attention while rinsing not to lose any of your polymer. It is very brittle and easily breaks into small pieces. Also, it is very thin, light and transparent making it harder to see when covered with water.

5. Nafion® modified with smaller salts dissolve easily in absolute ethanol while the larger ones are too hydrophobic and need more hydrophobic alcohol to completely dissolve. Tetrapropylammonium bromide (T3A), tetrabutylammonium bromide (TBAB), tetrapentylammonium bromide (T5A) and triethylhexylammonium bromide (TEHA) easily dissolve in absolute ethanol. Trimethylhexylammonium bromide (TMHA), trimethyloctylammonium bromide (TMOA), trimethyldecylammonium bromide (TMDA), trimethyldodecylammonium bromide (TMDDA), trimethyltetradecylammonium bromide (TMTDA), trimethylhexadecylammonium bromide (TMHDA), trimethyloctadecylammonium bromide (TMODA), and trimethylicosylammonium bromide (TMICA) dissolve in absolute propanol.

6. You can also store the modified polymer as solid flakes in capped and sealed glass vials for longer term stability.

7. Methanol is added to resuspend the gel and transform the polymer into a less viscous solution so further synthesis could take place homogeneously.

8. Reductive amination of chitosan with aldehyde is a very fast spontaneous exothermic reaction especially with aldehydes with smaller number of carbons. Fast addition of aldehyde will not form proper micelles within the polymer nor will they be evenly distributed due to instantaneous reaction of aldehyde with the amine group. Therefore, in order to form proper size and shape micelles homogeneously throughout the polymer, the aldehyde needs to be added very slowly and carefully. Because the reaction is very fast, the formation of intermediate polymer forms instantaneously seen by the clumping of solution which becomes impossible to stir with additions of aldehyde in large amounts and further preventing the proper homogeneous amination.

9. The addition of cyanoborohydride will transfer intermediately into a more stable product via the removal of the double bond between the aldehyde and the amine group.

10. If you are not doing a quantitative analysis, a faster way to obtain and filter is without any paper filter because most particles clump up into larger pieces.

11. It can be also dried at room temperature in a vacuum desiccator that is ran for an hour and then leave it over night to continually air dry.

12. Choose an alcohol which the enzyme is most tolerant, as well as a solvent the polymer will dissolve in. All enzymes mentioned here will tolerate both ethanol and propanol without a significant problem. However, the issue comes when the polymer does not completely dissolve in the chosen solvent. Ethanol has the capability to dissolve Nafion® that has been modified with less hydrophobic smaller salts. TMOA and TMHA modified Nafion® do dissolve in ethanol to some extent; however, they completely dissolve with propanol. Also the TMHDA, TMTDA, and TMODA modified Nafion polymers dissolve in propanol.

13. The enzyme stability within the solvent determines the time of mixing as well as the ratio of polymer to the enzyme solution.

14. Ensure that the electrodes are protected from collecting dust and other airborne particles that might interfere with proper micelle formation and enzyme encapsulation process. Double half closed boxes are efficient enough to minimize the contamination. Also keep the whole setup above or close to dehumidifier.

References

1. McKee T, McKee JR (eds) (1999) Biochemistry: an introduction, 2nd edn. The McGraw-Hill Companies, Inc., Boston, MA

2. Ragoonanan V, Aksan A (2007) Protein stabilization. Transfus Med Hemother 34:246–252

3. Palmore G, Bertschy H, Bergens SH, Whitesides GM (1998) A methanol/dioxygen biofuel cell that uses NAD+-dependent dehydrogenases as catalysts: application of an electro-enzymic method to regenerate nicotinamide adenine dinucleotide at low overpotentials. J Electroanal Chem 443(1):155–161

4. Mano N, Mao F, Heller A (2003) Characteristics of a miniature compartment-less glucose-O2 biofuel cell and its operation in a living plant. J Am Chem Soc 125(21):6588–6594

5. Mano N, Kim H, Heller A (2002) On the relationship between the characteristics of bilirubin oxidases and O$_2$ cathodes based on their "wiring". J Phys Chem B 106(34):8842–8848

6. Mano N, Mao F, Shin W, Chen T, Heller A (2003) A miniature biofuel cell operating at 0.78V. Chem Commun 4:518–519

7. Schindler JG, Schindler MM, Herna K, Pohl M, Guntermann H, Burk B, Reisinger E (1994) Long-functioning beta-D-glucose and L-lactate biosensors for continuous flow-through measurements for "fouling"-resistant and selectivity-optimized serum- and hemo-analysis. Eur J Clin Chem Clin Biochem 32(8):599–608

8. Treu BL (2008) Development of DET-based PQQ-dependent enzymatic biofuel cells. Saint Louis University, St. Louis

9. Kim E, Kim K, Yang H, Kim YT, Kwak J (2003) Enzyme-amplified electrochemical detection of DNA using electrocatalysis of ferrocenyl-tethered dendrimer. Anal Chem 75:5665–5672

10. Calabrese-Barton S, Kim H, Binyamin G, Zhang Y, Heller A (2001) Electroreduction of O$_2$ to water on the "wired" laccase cathode. J Phys Chem B 105(47):11917–11921

11. Mano N, Mao F, Heller A (2002) A miniature biofuel cell operating in a physiological buffer. J Am Chem Soc 124(44):12962–12963

12. Kim H, Zhang Y, Heller A (2004) Bilirubin oxidase label for an enzyme-linked affinity assay with O$_2$ as substrate in a neutral pH NaCl solution. Anal Chem 76(8):2411–2414

13. Wu Y, Zheng Y, Yang W, Wang C, Hu J, Fu S (2005) Synthesis and characterization of a novel amphiphilic chitosan–polylactide graft copolymer. Carbohydrate Polymer 59:165–171

14. Kataoka K, Kwon GS, Yokoyama M, Okano T, Sakurai Y (1993) Block copolymer micelles as vehicles for drug delivery. J Control Release 24:119–132

15. Gierke TD, Hsu WY (eds) (1982) Perflourinated Ionomer Membranes: The cluster-network model of ion clustering in perfluorosulfonated membranes. ACS Symposium Series 180, American Chemical Society, Washington, DC

16. Schrenk MJ, Villigram RE, Torrence NJ, Brancato SJ, Minteer SD (2002) Effects of mixture casting Nafion® with quaternary ammonium bromide salts on the ion-exchange capacity and mass transport in the membranes. J Membr Sci 205:3–10

17. Menius CE, Arechederra RL, Minteer SD (2009) Microscopic characterization of micellar structures in modified Nafion films. PMSE Prepr 100. pp 625–626

18. Giroud F, Minteer SD (2013) Anthracene-modified pyrenes immobilized on carbon nanotubes for direct electroreduction of O2 by laccase. Electrochem Commun 34:157–160. doi:10.1016/j.elecom.2013.06.006

19. Meredith MT, Minson M, Hickey D, Artyushkova K, Glatzhofer DT, Minteer SD (2011) Anthracene-modified multi-walled carbon nanotubes as direct electron transfer scaffolds for enzymatic oxygen reduction. ACS Catal 1:1683–1690. doi:10.1021/cs200475q

20. Thomas TJ, Ponnusamy KE, Chang NM, Galmore K, Minteer SD (2003) Effects of annealing on mixture-cast membranes of Nafion® and quaternary ammonium bromide salts. J Membr Sci 213:55–66

21. Klotzbach TL, Watt MM, Ansari Y, Minteer SD (2006) Effects of hydrophobic modification of Chitosan and Nafion on transport properties, ion-exchange capacities, and enzyme immobilization. J Membrane Sci 282(1):276–283

22. Moore CM, Akers NL, Hill AD, Johnson ZC, Minteer SD (2004) Improving the environment for immobilized dehydrogenase enzymes by modifying nafion with tetraalkylammonium bromides. Biomacromolecules 5(4):1241–1247

23. Shjak-Braek G, Anthonsen T, Sandford P (1988) Chitin and Chitosan: sources, chemistry, biochemistry, physical properties and applications. Elsevier

24. Batrakova EV, Han H-Y, Alakhov VY, Miller DW, Kabanov AV (1998) Effects of pluronic block copolymers on drug absorption in Caco-2 cell monolayers. Pharm Res 15(6):850–855

25. Gerente C, Lee VKC, Cloirec PL, McKay G (2007) Applications of Chitosan for the removal of metals from wastewaters by adsorption-mechanism and models review. Crit Rev Environ Sci Technol 37:41–127, doi: 1547-6537 online

26. Esquenet C, Buhler E (2001) Phase behavior of associating polyelectrolyte polysaccharides. 1. Aggregation process in dilute solution. Macromolecules 34(15):5287–5294

27. Kurita K (2006) Chitin and Chitosan: functional biopolymers from marine crustaceans. Marine Biotechnol 8(3):203–226

28. Beaulieu C (2007) The multiple effects of chitin. Gen Rev Phytothérapie HS38–HS45

29. Yalpani M, Hall L (1984) Some chemical and analytical aspects of polysaccharide modifications. 3. Formation of branched-chain, soluble chitosan derivatives. Macromolecules 17(3):272–281

30. Jinrui X, Bin L (1994) Preconcentration and determination of lead ions at a chitosan-modified glassy carbon electrode. Analyst 119:1599–1601

31. Wu X, Lu G, Zhang X, Yao X (2001) Anodic voltammetric behavior of iodide at a chitosan-modified glassy carbon electrode. Anal Lett 34:1205–1214

32. Xin Y, Guanghan L, Xiaogang W, Tong Z (2001) Studies on electrochemical behavior of bromide at a chitosan-modified glassy carbon electrode. Electroanalysis 13:923–926

33. Ye X, Yang Q, Wang Y, Li N (1998) Electrochemical behaviour of gold, silver, platinum and palladium on the glassy carbon electrode modified by chitosan and its application. Talanta 47:1099–1106

34. Huang H, Hu N, Zeng Y, Zhou G (2002) Electrochemistry and electrocatalysis with heme proteins in chitosan biopolymer films. Anal Biochem 308(1):141–151

35. Hikima S, Kakizaki T, Taga M, Hasebe K (1993) Enzyme sensor for L-lactate with chitosan-mercury film electrode. Fresenius J Anal Chem 345:607–609

36. Miao Y, Tan SN (2000) Amperometric hydrogen peroxide biosensor based on immobilization of peroxidase in chitosan matrix crosslinked with glutaraldehyde. Analyst 125:1591–1594

37. Esquenet C, Terech P, Boue F (2004) Structural and rheological properties of hydrophobically modified polysaccharide associative networks. Langmuir 20(9):3583–3592

38. Desbrieres J, Rinaudo M, Chtcheglova L (1997) Reversible thermothickening of aqueous solutions of polycations from natural origin. Macromol Symp 113:135–149

39. Klotzbach TL, Watt MM, Ansari Y, Minteer SD (2008) Improving the microenvironment for enzyme immobilization at electrodes by hydrophobically modifying Chitosan and Nafion polymers. J Membr Sci 311:81–88

Chapter 9

Cross-Linked Enzyme Aggregates for Applications in Aqueous and Nonaqueous Media

Ipsita Roy, Joyeeta Mukherjee, and Munishwar N. Gupta

Abstract

Extensive cross-linking of a precipitate of a protein by a cross-linking reagent (glutaraldehyde has been most commonly used) creates an insoluble enzyme preparation called cross-linked enzyme aggregates (CLEAs). CLEAs show high stability and performance in conventional aqueous as well as nonaqueous media. These are also stable at fairly high temperatures. CLEAs with more than one kind of enzyme activity can be prepared, and such CLEAs are called combi-CLEAs or multipurpose CLEAs. Extent of cross-linking often influences their morphology, stability, activity, and enantioselectivity.

Key words Cross-linking agents, Glutaraldehyde, Cross-linked enzyme aggregates, Immobilization, Thermal stability, Enantioselectivity

1 Introduction

The cross-linking techniques (that is, use of bifunctional reagents) for proteins were introduced by Wold's group. Wold also wrote an excellent review on the technique [1] which he updated again later on [2]. The reaction of a protein with a bifunctional reagent [1, 2] may give rise to a variety of products: (a) If only one side chain on the protein reacts, this leads to monofunctional modification of the protein just as in the case of a monofunctional reagent. (b) The reagent forms an intramolecular cross-link in the protein, a possibility favored at low concentration of the protein. (c) The reagent forms an intermolecular cross-link between different molecules of

---

Prof. Finn Wold, while at University of Minnesota, St. Paul, USA, introduced bifunctional reagents (more frequently called cross-linking reagents) to protein chemistry. Consequently several subsequent developments including CLEA design were possible. Prof. Wold was one of the early mentors of one of the authors (Munishwar N. Gupta). This chapter is dedicated to the memory of Prof. Finn Wold who was a great scientist and a great human being.

Shelley D. Minteer (ed.), *Enzyme Stabilization and Immobilization: Methods and Protocols*, Methods in Molecular Biology, vol. 1504, DOI 10.1007/978-1-4939-6499-4_9, © Springer Science+Business Media New York 2017

the protein to form soluble oligomers of the protein. (d) When high concentration of both protein and the reagent are used, extensive intermolecular cross-linking (accompanied by intramolecular cross-linking to a varying degree depending upon the enzyme) leads to formation of insoluble chemical aggregates. A large number of such chemical aggregates have been described in the literature [3]. In early reviews [3, 4], these chemical aggregates were considered as immobilized enzymes without a matrix. A somewhat different way of forming such insoluble aggregates was described by Sheldon's group. In this version, the protein is precipitated by a salt/organic solvent and cross-linked while still in the precipitated form. Such insoluble aggregates have been called cross-linked enzyme aggregates (CLEAs) [5–8]. CLEAs are said to be less amorphous, easier to handle, show high stability at high temperatures, in the presence of organic solvents and even in nearly anhydrous organic solvents [7, 9, 10]. CLEA preparations also show high reusability. CLEA of (R)-oxynitrilase could be recycled ten times without loss of activity [9]. Majumder et al. [10] have shown the increase in half-life of CLEA of *Burkholderia cepacia* lipase (BCL) at 55 °C in an aqueous buffer. Similarly, half-lives of pectinase (at 50 °C), xylanase (at 60 °C), and cellulase (at 70 °C) improved significantly upon formation of a multipurpose CLEA [11]. There are a few examples wherein CLEA prepared in the presence of additives showed improved thermal stability [12]. There are also a large number of examples wherein CLEAs have shown much higher activity in organic media. Sheldon [13] mentioned that CLEA of *Candida antarctica* lipase B (CALB) showed six times higher activity (in diisopropyl ether) than the commercially available immobilized form, Novozym 435. In another example, CLEA of penicillin acylase performed much better (than even carrier-bound enzyme) in kinetically controlled synthesis of ampicillin in 60 % (v/v) ethylene glycol [14]. CLEA of (R)-oxynitrilase worked as an efficient hydrocyanation catalyst under microaqueous conditions [9]. For further examples, many good reviews/chapters can be consulted [8, 13, 15]. CLEAs are also reported to show good performance in ionic liquids [16, 17]. Table 1 shows that CLEAs of both *Candida rugosa* lipase (CRL) as well as BCL performed quite well in the kinetic resolution of 1-phenylethanol in [Bmim][PF$_6$] (1-butyl-3-methylimidazolium hexafluorophosphate). The table also compares the performance of these CLEAs with the pH tuned lyophilized powders and EPROS (enzymes precipitated and rinsed with organic solvents; precipitates of enzymes obtained by the optimal use of an appropriate organic solvent) [18–21]. It is interesting to note that these EPROS preparations performed quite well as compared to the corresponding CLEAs. In fact, in case of CRL, EPROS was better than the corresponding CLEA. CLEAs have also been used in supercritical CO$_2$ [22].

Table 1
Kinetic resolution of 1-phenylethanol in [Bmim][PF$_6$] catalyzed by different preparations of lipases

Entry	Lipase preparation	Time (h)	Conversion (%)	ee$_p$ (%)[a]	ee$_s$ (%)[a]	E[b]
Candida rugosa lipase						
1	Lyophilized powder	12	5	80	5	11
2	Lyophilized powder	24	7	83	6	12
3	EPROS	12	22	97	28	123
4	EPROS	24	26	98	34	153
5	CLEA	12	10	91	10	30
6	CLEA	24	12	95	12	69
Burkholderia cepacia lipase						
7	Lyophilized powder	1	5	99	5	114
8	Lyophilized powder	2	8	99	8	187
9	EPROS	1	31	99	45	314
10	EPROS	2	49	99	96	736
11	CLEA	1	41	99	69	432
12	CLEA	2	50	99	99	>1000

Adapted from ref. [17] with permission from Elsevier
[a]Conversion and ee's were estimated by HPLC
[b]Enantioselectivity (E) was calculated according to Chen's equation [17]

Obtaining chirally pure compounds is another valuable application of biocatalysts. CLEA have shown improved enantioselectivity in a few cases. A good example is synthesis of enantiomerically pure (S)-mandelic acid from benzaldehyde using the bienzymatic cascade of CLEA comprising (S)-selective oxynitrilase from *Manihot esculenta* and the nonselective recombinant nitrilase from *Pseudomonas fluorescens* EBC 191 [23].

CLEAs are often compared with a prior biocatalyst design called cross-linked enzyme crystals (CLECs). CLECs are obtained by cross-linking of microcrystals of protein and unlike CLEAs, require fairly pure protein preparations [24].

The possibility of formation of CLEAs with mixed population of enzyme molecules also enables one to make combi-CLEAs (with which a sequence of two successive reactions can be carried out by CLEAs formed from a mixture of two enzymes) [23] or multipurpose CLEAs (CLEA formed from mixtures of enzymes with quite different and unrelated activities such as the CLEA containing pectinase, xylanase, and cellulase activities) [11]. In many studies, CLEAs have been shown to perform better than just lyophilized

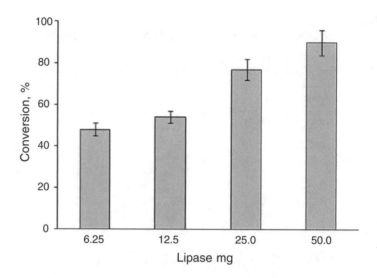

Fig. 1 Biodiesel production from Mahua oil catalyzed by CLEAs of *Pseudomonas cepacia* (*Burkholderia cepacia*) lipase. CLEAs were prepared from different amounts of lipase and added to the reaction media containing Mahua oil and ethanol at a ratio of 1:4. These mixtures were incubated for 2.5 h at 40 °C with a constant shaking at 200 rpm. The experiment was performed in triplicate, and the error bars represent the percentage error in each set of readings (Adapted from ref. [25] with permission from American Chemical Society, Copyright 2007)

powders in organic media [12, 17]. Control of extent of cross-linking during CLEA preparation makes it possible to choose the desired trade-off between activity, stability and enantioselectivity [10]. The CLEA of BCL was also found to catalyze efficient production of biodiesel from Mahua oil, which contains a high amount of free fatty acids ([25], Fig. 1).

Some recent applications of CLEAs as a biocatalyst include hydrolysis of narangin [26]; enrichment of PUFA (polyunsaturated fatty acids) in fish oil by hydrolysis catalyzed by Geotrichium sp. lipase [27]; synthesis of L-phenylalanine using CLEA of phenylalanine ammonia lyase from *Rhodotorula glutinis* [28]; papain-catalyzed hydrolysis of BSA and ovalbumin [29]; hydrolysis of poly-3-hydroxybutyrate by CLEA of the depolymerase from *Streptomyces exfoliatus* [30]; stabilization of CRL against transacetylation so that the lipase is not vulnerable to acetaldehyde generated in situ in the reaction of vinyl acetate with benzyl alcohols [31]. Guauque et al. [32] have described a procedure for correctly calculating the activity of the CLEAs. CLEAs of beta-galactosidase have been used in a biphasic system to avoid substrate inhibition [33].

With multimeric enzymes, CLEA methodology offers the advantage of stabilization of quaternary structures. An example is the tetrameric catalases [34]. While CLEAs are called carrier free immobilized enzymes, some variations on the methodology involves matrices. For specific applications, such innovations may be useful. Kim et al. [35] have entrapped CLEAs of α-chymotrypsin and *Mucor javanicus* lipase in pores of hierarchically ordered mesoporous silica. Wilson et al. [34] have encapsulated penicillin G acylase CLEA in LentiKats. Hilal et al. [36] have entrapped lipase CLEA within microporous polymeric membranes. Table 2 provides a snapshot of the different approaches used over the years for

Table 2
Types of enzyme aggregates

Type of aggregate	Unique features	Reference
Insoluble amorphous "chemical aggregates"	These aggregates often retain fairly high activity and show somewhat enhanced stability under denaturing conditions such as moderately high temperature and organic cosolvents	[40]
Cross-linked enzyme aggregates (CLEA)	In this approach, protein is cross-linked when present in precipitate form. This strategy has resulted in remarkable improvement in the following properties of the resultant biocatalyst: (a) Higher retention of biological activity; 100% in a large number of cases; (b) Better reproducibility by easier control of parameters. Chemical aggregate formation requires a delicate balance between several factors, such as the amount of cross-linker, temperature, pH, and ionic strength; (c) Higher mechanical stability and stability towards denaturing conditions	[10, 12]
CLEA with polymeric additive	Formed by co-precipitation of the enzyme with poly-ethyleneimine (PEI) or PEI–dextran sulfate mixtures. This produces significant changes in the activity, specificity, and enantioselectivity of the enzyme. This approach can be used for the preparation of CLEAs of enzymes with low number of lysines on their surface	[34, 41]
CLEA with bovine serum albumin (BSA) as proteic feeder	Addition of BSA facilitates obtaining CLEAs in cases where the protein concentration in the enzyme preparation is low and/or the enzyme activity is vulnerable to the high concentration of glutaraldehyde required to obtain aggregates	[12]
Cross-linked imprinted aggregates (CLIP)	Different precipitants result in different enzyme conformations which are "frozen" by extensive cross-linking. This interesting approach is called cross-linked imprinting (CLIP) approach	[42]
Combi-CLEA	The approach of combi-CLEA is aimed at turning CLEA into a multi-enzyme complex for multistep conversions for cascade catalysis	[23]

(continued)

Table 2
(continued)

Type of aggregate	Unique features	Reference
Multipurpose CLEA	In this approach, the CLEA may contain several enzyme activities for catalyzing non-cascade reactions	[43]
Laccase CLEAs using chitosan	While the authors described chitosan as a novel cross-linker, it was actually used as a spacer to be inserted between carboxyl groups of the enzyme via carbodiimide coupling. The thermal stability improved but stability against chemical denaturants did not	[44]
Use of recombinant *V. spinosum* tyrosinase as the cross-linker	Addition of phenol to compact globular proteins like lysozyme and CAL B facilitated formation of CLEAs by addition of tyrosinase	[45]
Fast PREP CLEAs	*N*-acetyl muramic acid-aldolase CLEAs on dispersion by a reciprocating disrupter (Fast PREP) produced a twofold increase in aldolase activity	[46]
Porous CLEAs of papain	Starch was added as a pore making agent and was hydrolysed subsequently by alpha-amylase. Tis resulted in improved activities	[29]
CLEAs for synthesis in CO_2-expanded micellar solutions	Trypsin was precipitated from the CO_2-expanded reverse micellar solutions and then cross-linked to give dendritic CLEAs. The activity of CLEAs obtained by this method is improved in contrast to those obtained by the conventional method	[47]
Hybrid CLEA of phenylalanine ammonium lyase from *Rhodotorula glutinis*	Novel hybrid magnetic cross-linked enzyme aggregates of phenylalanine ammonia were developed by co-aggregation of enzyme aggregates with magnetite nanoparticles and subsequent cross-linking with glutaraldehyde. The HM-PAL-CLEAs can be easily separated from the reaction mixture by using an external magnetic field	[48]
CLEA of phenylalanine ammonium lyase in microporous silica gel (MSG-CLEAs)	MSG-CLEAs exhibited the excellent stability of the enzyme against various deactivating conditions such as temperature and denaturants, and showed higher storage stability compared to the free PAL and the conventional CLEAs of the same enzyme	[49]
Magnetic CLEAs of alpha amylase	These were prepared by chemical cross-linking of enzyme aggregates with amino functionalized magnetite nanoparticles which can be separated from reaction mixture using magnetic field. 100 % of the initial activity was recovered in magnetic CLEAs, whereas only 45 % was recovered in CLEAs due to the low content of Lys residues in alpha amylase	[50]

(continued)

Table 2
(continued)

Type of aggregate	Unique features	Reference
CLEAs entrapped in 3D ordered macroporous silica	The lipase was first precipitated in the pores of the silica using ammonium sulfate and then cross-linked by glutaraldehyde. These CLEAs exhibited excellent thermal and mechanical stability, and could maintain more than 85 % of initial activity after 16 days of shaking in organic and aqueous phase	[51]
Combi CLEAs of a ketoreductase and D-glucose dehydrogenase	Cofactor regeneration in the synthesis of valuable chiral alcohols catalyzed by ketoreductases was more efficiently possible with these CLEAs. combi-CLEAs of ketoreductase and D-glucose dehydrogenase enabled the repeated and effective conversion of substrate ethyl 4-chloro-3-oxobutanoate (COBE)	[52]
Combi CLEAs of amylosucrase (AS), maltooligosyltrehalose synthase (MTS), and maltooligosyltrehalose trehalohydrolase (MTH)	CLEA-AS-MTS-MTH was used to convert sucrose to trehalose	[53]
Hybrid bioreactor of hollow fiber membrane and CLEAs of laccase	Used for effluent wastewater treatment for the removal of aromatic pharmaceuticals	[54]

obtaining enzyme aggregates for catalytic purposes. It is worth noting that with growing popularity of the CLEA concept, the line between amorphous aggregates and patented procedure for CLEAs has been blurred. All chemically cross-linked aggregates are being called CLEAs! It is also interesting to note that recent work has suggested the use of inclusion bodies (IBs) (with significant level of catalytic activity) in lieu of CLEAs [37].

This chapter describes:

Protocol 3.1: A general method for the preparation of CLEAs.

Protocol 3.2: Preparation of CLEAs with BSA as a proteic feeder and illustrated with the preparation of Penicillin acylase CLEA.

Protocol 3.3: Preparation of multipurpose CLEAs with pectinase, xylanase, and cellulose additives.

2   Materials

1. 3-Å molecular sieves (E. Merck, Mumbai, India).

2. Glutaraldehyde solution: 25 %, v/v in water.

3. Pasteur pipettes.

4. All solvents used are of low water grade and are further dried by gentle shaking with 3-Å molecular sieves.

5. Sodium phosphate buffer: 50 mM and pH 7.0.

6. Tris–HCl buffer: 100 mM and pH 8.0.

7. Sodium acetate buffer: 50 mM and pH 5.0.

8. *Candida rugosa* lipase (Amano Enzyme Inc., Nagoya, Japan).

9. *Pseudomonas cepacia* (*Burkholderia cepacia*) lipase (Amano Enzyme Inc., Nagoya, Japan).

10. Penicillin G acylase (Fluka) (It is a liquid enzyme).

11. Pectinex™ Ultra SP-L (a commercial preparation of pectolytic enzymes from a selected strain of *Aspergillus niger*) (Novozymes, Bangalore, India).

3 Methods

The exact conditions for the preparation of CLEAs vary from enzyme to enzyme. Thus, the conditions should first be optimized with respect to a particular enzyme in terms of: (a) choice of precipitant which does not lead to loss of enzyme activity and precipitates the enzyme quantitatively. Some of the precipitants which have worked well are: acetone, dimethoxyethane, n-propanol, t-butanol, polyethylene glycol (PEG), and ammonium sulfate; (b) requirement of an additive which can be either a synthetic polymer or a proteic feeder. Addition of the inert protein (e.g., bovine serum albumin, ovalbumin, hemoglobin, or fibrinogen), called proteic feeder, prevented reaction with amino groups located at or near the enzyme active site and did not allow excessive cross-linking which results in too rigid a conformation of the enzyme. This over-rigidity led to poorly active enzyme molecules. The same concept of proteic feeder works equally well in the case of CLEA formation. It was shown that addition of BSA before precipitation and cross-linking resulted in (1) facilitating required level of cross-linking in the case of a *Pseudomonas cepacia* lipase preparation which had very poor protein content [8] and (2) a more active CLEA of penicillin G acylase as otherwise some essential amino groups of the enzyme were modified by glutaraldehyde [12]. The morphologies of CLEAs obtained with or without the addition of BSA in the case of penicillin G acylase were different as seen by scanning electron microscopy (SEM) and atomic force microscopy (AFM) (Fig. 2); (c) nature of cross-linking agent. In a majority of the cases, glutaraldehyde has been used for cross-linking. In some cases, wherein reaction with glutaraldehyde led to loss in enzymatic activity, a larger molecule, dextran polyaldehyde, has been reported to be a better choice as the reaction becomes limited to more accessible amino groups present on the protein surface [38]; (d) cross-linker concentration; and (e) cross-linking time.

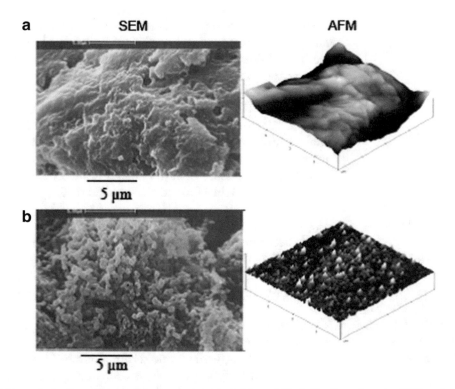

Fig. 2 Characterization of cross-linked enzyme aggregates of Penicillin G acylase by SEM and AFM. (**a**) Scanning electron micrograph and atomic force micrograph of CLEA without BSA. (**b**) Scanning electron micrograph and atomic force micrograph of CLEA with BSA (Adapted from ref. [12] with permission from Elsevier)

3.1 Preparation of CLEAs

1. The prechilled precipitant (4 times volume of the aqueous solution) is taken in a capped centrifuge tube and the tube is then kept on an orbital shaker (set at 150 rpm, 4 °C).

2. The enzyme solution made in 50 mM phosphate buffer, pH 7.0 (*see* **Notes 1** and **2**) is added dropwise to the precipitant (*see* **Note 3**) for protein aggregation by simultaneously shaking the tube.

3. After 30 min, the physically aggregated enzyme is subjected to chemical cross-linking using an optimized concentration of glutaraldehyde (which varies from 5 to 150 mM) under constant shaking condition (*see* **Notes 4** and **5**). After the addition of glutaraldehyde, the mixture in the tube is shaken at 300 rpm at 4 °C for 3–4 h (*see* **Note 6**).

4. The reaction may be stopped by the addition of 100 mM Tris buffer, pH 8.0. In our experience, this step may be skipped since the subsequent centrifugation and washing steps anyway separate the CLEAs from the unreacted glutaraldehyde [10–12, 17].

5. The tube is centrifuged at $8000 \times g$ at 4 °C for 10 min to separate the CLEAs. The supernatant is decanted, and the CLEAs are washed three to four times with the buffer (50 mM phosphate

buffer, pH 7.0) by repeatedly suspending the CLEAs in buffer and centrifuging to separate the CLEAs (*see* **Note 7**). The washed CLEAs are finally suspended in the assay buffer and used for enzyme assays. For applications in low water media, CLEAs should be washed with a water miscible organic solvent two to three times. Finally, these should be rinsed with the desired organic solvent (reaction medium) and incubated with the same solvent equilibrated with the desired water activity (water activity, a_w, is given by the ratio of the partial pressure of water in the system to the pressure of pure water) or water content [10, 17, 39]. The CLEAs thus formed are stored at 4 °C.

3.2 Preparation of CLEAs with Additives

CLEAs with polymeric additives or BSA as a proteic feeder are prepared according to the same procedure as described in Subheading 3.1, but before adding the precipitant to the enzyme solution, polymeric additive or proteic feeder is added to the enzyme solution under stirring conditions for 10 min [12].

3.2.1 Preparation of CLEAs of Lipase Using BSA as a Proteic Feeder

1. *Candida rugosa* lipase (CRL) or *Burkholderia cepacia* lipase (BCL) (10 mg) is dissolved in 1 ml of 50 mM phosphate buffer, pH 7.0 along with bovine serum albumin (2.5 mg). This solution is added dropwise to a centrifuge tube containing chilled acetone (4 ml) at 4 °C.

2. After 30 min, glutaraldehyde is added up to a final concentration of 25 mM. The mixture is incubated at 4 °C for 3 h at 300 rpm (constant shaking).

3. After 3 h, the solution is centrifuged at $8000 \times g$ for 10 min at 4 °C. The CLEAs formed are washed three times with 1 ml of acetone and finally stored in acetone.

4. Prior to use, CLEAs are washed twice with 1 ml of cold reaction medium (*see* **Note 8**).

3.2.2 Preparation of Penicillin Acylase CLEAs Using BSA as a Proteic Feeder

1. Penicillin G acylase (125 µl) containing 5 mg of BSA is added dropwise to a shaking centrifuge tube (150 rpm) containing 400 µl of dimethoxyethane at 4 °C.

2. After 30 min, glutaraldehyde is added so that the final concentration of glutaraldehyde is 50 mM. The mixture is incubated at 4 °C for 4 h with constant shaking at 300 rpm.

3. After 4 h, dimethoxyethane (1 ml) is added to this mixture, which is then centrifuged at $8000 \times g$ at 4 °C. The supernatant is decanted, and the CLEAs are washed three times with 50 mM phosphate buffer, pH 7.0 (5 ml). The CLEAs are finally stored in the same buffer at 4 °C and used for further characterization (*see* Fig. 2 and **Note 9**).

3.3 Preparation of Multipurpose CLEAs Containing Pectinase, Xylanase, and Cellulase Activities

1. Pectinex™ Ultra SP-L (crude enzyme solution containing pectinase, xylanase, and cellulase activities) is used for the preparation of multipurpose CLEAs [11].

2. Chilled n-propanol (5 ml) is added dropwise to crude enzyme solution (1 ml) in a shaking (150 rpm) capped centrifuge tube at 4 °C.

3. After 15 min, glutaraldehyde is added up to a final concentration of 5 mM. The mixture is incubated at 4 °C for 4 h with constant shaking at 300 rpm.

4. After 4 h, the tube is centrifuged at $10,000 \times g$ at 4 °C for 5 min. The supernatant is decanted, and the CLEAs are washed three times with the 50 mM sodium acetate buffer, pH 5.0 (5 ml). The CLEAs are finally stored in the same buffer (1 ml) at 4 °C.

4 Notes

1. Unless stated otherwise, all solutions should be prepared in water that has a resistivity of 18.2 MΩ cm and total organic content of less than five parts per billion.

2. The use of Tris buffer or any amine buffer should be avoided for the preparation of CLEAs since amine buffers interfere with the cross-linking reaction. The protein concentration in the solution should be ≥2 mg/ml.

3. In a majority of the cases, we have found precipitation of enzymes with dimethoxyethane (DME) and acetone resulted in retention of 100 % activity. In case where the cell lysates are directly used for the preparation of CLEAs, precipitation with ammonium sulfate was preferred since it resulted in the simultaneous removal of nucleic acids in the supernatant. While using ammonium sulfate as the precipitant, ammonium sulfate was slowly added to the enzyme solution in the centrifuge tube under constant shaking.

4. Glutaraldehyde should be stored between 2 and 8 °C. When handling glutaraldehyde-based solutions, avoid contact with the liquid and inhalation of the vapor. Protective gloves, splash-proof monogoggles, and protective clothing should be worn. In most of the cases, glutaraldehyde has been used for cross-linking but similar precautions may have to be taken while using other cross-linking agents.

5. It was found that highly cross-linked aggregates were morphologically different and exhibited better thermal stability. Such cross-linked aggregates though showed poor enantioselectivity [10]. X-ray diffraction pattern reveals that cross-linking caused a drop in crystallinity and an increase in amorphous character in CLEAs [10]. Schoevaart et al. [7] indicated that morphologically CLEAs have either a ball-like appearance (type 1) or a

less-structured form (type 2). This was said to depend upon the nature of the enzyme. However, in a later work it was shown that same enzyme may form CLEAs of different appearances depending upon the extent of cross-linking [10].

6. The cross-linking time may vary from 10 min to 5 h depending on the enzyme to be cross-linked. Thus, the cross-linking time should be optimized for the preparation of CLEAs in the individual cases.

7. After the cross-linking reaction is over, the CLEAs should be washed thoroughly with buffer to remove the unreacted cross-linking reagent. If this step is not done properly, it may interfere with the enzyme assay of the cross-linked preparation.

8. This step is carried out to remove the traces of the precipitating agent and is especially critical when CLEAs are used in organic solvents. Sometimes the final wash may also be with one of the substrates, for example, with chilled vinyl acetate in case of the kinetic resolution of (±)-1-phenylethanol in ionic liquids [17]. Ionic liquids such as [Bmim][PF$_6$] are expensive and hence washing with one of the substrates is a better option.

9. The CLEAs prepared in the presence of BSA as the proteic feeder are also found to be thermally more stable (Fig. 3).

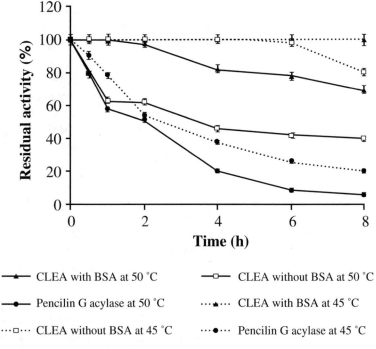

Fig. 3 Comparison of thermal stability of penicillin G acylase CLEA prepared in the presence and absence of BSA. Thermal stability was studied at 45 and 50 °C in 100 mM phosphate buffer, pH 8 (Reproduced from ref. [12] with permission from Elsevier Science)

For example, at 45 °C, CLEAs of penicillin G acylase prepared in the presence of BSA retained 100% activity up to 8 h. Even at 50 °C, CLEAs prepared in the presence of BSA were significantly more stable. It may be mentioned that CLEAs of penicillin G acylase prepared in the presence of ionic polymers did not show any thermal stabilization as compared to the standard CLEA [33]. On the other hand, CLEAs prepared in the presence of BSA showed higher operational stability in organic solvents (in the case of lipase) and higher thermal stability (in the case of penicillin G acylase) [12].

Acknowledgments

This work was supported by funds obtained from Department of Science and Technology [Grant No.: SR/SO/BB-68/2010] and Department of Biotechnology [Grant No.: BT/PR14103/BRB/10/808/2010], both Government of India organizations. Finally, we thank past members of our research group; Dr. Kalyani Mondal, Dr. Shweta Shah, Dr. Abir Majumder, Dr. Sohel Dalal, and Veena Singh, whose work has been described/quoted in this chapter.

References

1. Wold F (1967) Bifunctional reagents. Methods Enzymol 11:617–640
2. Wold F (1972) Bifunctional reagents. Methods Enzymol 25:623–651
3. Broun GB (1976) Chemically aggregated enzymes. Methods Enzymol 44:263–280
4. Gupta MN (1993) Applications of crosslinking techniques to enzyme/protein stabilization and bioconjugate preparation. In: Himmel ME, Georgiou G (eds) Biocatalyst design for stability and specificity. ACS Symposium Series Am. Chem. Soc, Washington, DC, pp 307–324
5. Cao L, van Rantwijk F, Sheldon RA (2000) Cross-linked enzyme aggregates: a simple and effective method for the immobilization of penicillin acylase. Org Lett 2:1361–1364
6. Cao L, van Langen LM, van Rantwijk F, Sheldon RA (2001) Crosslinked aggregates of penicillin acylase. Robust catalyst for the synthesis of ß-lactam antibiotics. J Mol Catal B Enzym 11:665–670
7. Schoevaart R, Wolbers MW, Golubovic M, Ottens M, Kieboom AP, van Rantwijk F, van der Wielen LA, Sheldon RA (2004) Preparation, optimization, and structures of cross-linked enzyme aggregates (CLEAs). Biotechnol Bioeng 87:754–762
8. Sheldon RA, Schoevaart R, van Langen LM (2006) Cross-linked enzyme aggregates. In: Guisan JM (ed) Immobilization of enzymes and cells. Humana Press, Totowa, NJ, p 43
9. van Langen LM, Selassa RP, van Rantwijk F, Sheldon RA (2005) Cross-linked aggregates of (R)-oxynitrilase: a stable, recyclable biocatalyst for enantioselective hydrocyanation. Org Lett 7:327–329
10. Majumder AB, Mondal K, Singh TP, Gupta MN (2008) Designing cross-linked lipase aggregates for optimum performance as biocatalysts. Biocatal Biotransform 26:235–242
11. Dalal S, Sharma A, Gupta MN (2007) A multipurpose immobilized biocatalyst with pectinase, xylanase and cellulase activities. Chem Cent J 1:16
12. Shah S, Sharma A, Gupta MN (2006) Preparation of cross-linked enzyme aggregates by using bovine serum albumin as a proteic feeder. Anal Biochem 351:207–213
13. Sheldon RA (2006) Immobilization of enzymes as cross-linked enzyme aggregates: a simple method for improving performance. In: Patel RN (ed) Biocatalysis in the pharmaceutical and biotechnology industries. CRC Press, Boca Raton, NY, pp 350–362
14. Illanes A, Wilson L, Caballero E, Fernández-Lafuente R, Guisan JM (2006) Cross-linked penicillin acylase aggregates for synthesis of β-lactam antibiotics in organic medium. Appl Biochem Biotechnol 133:189–202

15. Sheldon RA, Schoevaart R, van Landen LM (2005) Cross-linked enzyme aggregates (CLEAs): a novel and versatile method for enzyme immobilization (a review). Biocatal Biotransform 23:141–147

16. Ruiz Toral A, de Los Rios AP, Hernandez FJ, Janssen MHA, Schoevaart R, van Rantwijk F, Sheldon RA (2007) Cross-linked Candida antarctica lipase B is active in denaturing ionic liquids. Enzyme Microb Technol 40:1095–1099

17. Shah S, Gupta MN (2007) Kinetic resolution of (±)-1-phenylethanol in [Bmim][PF₆] using high activity preparations of lipases. Bioorg Med Chem Lett 17:921–924

18. Roy I, Gupta MN (2004) Preparation of highly active alpha-chymotrypsin for catalysis in organic media. Bioorg Med Chem Lett 14:2191–2193

19. Solanki K, Gupta MN (2008) Optimizing biocatalyst design for obtaining high transesterification activity by a-chymotrypsin in non-aqueous media. Chem Cent J 2:1–7

20. Majumder AB, Gupta MN (2011) Increasing catalytic efficiency of *Candida rugosa* lipase for the synthesis of tert-alkyl butyrates in low water media. Biocatal Biotrasform 29:238–245

21. Solanki K, Gupta MN, Halling PJ (2012) Examining structure-activity correlations of some high activity enzyme preparations for low water media. Bioresour Technol 115:147–151

22. Hobbs HR, Kondor B, Stephenson P, Sheldon RA, Thomas NR, Poliakoff M (2006) Continuous kinetic resolution catalysed by cross-linked enzyme aggregates, "CLEAs", in supercritical CO₂. Green Chem 8:816–821

23. Mateo B, Chmura A, Rustler S, van Rantwijk F, Stolz A, Sheldon RA (2006) Synthesis of enantiomerically pure (S)-mandelic acid using an oxynitrilase–nitrilase bienzymatic cascade: a nitrilase surprisingly shows nitrile hydratase activity. Tetrahedron Asymm 17:320–323

24. St. Clair NL, Navia MA (1992) Cross-linked enzyme crystal as robust biocatalysts. J Am Chem Soc 114:7314–7316

25. Kumari V, Shah S, Gupta MN (2007) Preparation of biodiesel by lipase-catalyzed transesterification of high free fatty acid containing oil from *Madhuca indica*. Energ Fuel 21:368–372

26. Ribero MH, Rabaca M (2011) Cross-linked enzyme aggregates of naringinase: novel biocatalysts for naringin hydrolysis. Enzym Res 2011:851272

27. Yan J, Gui X, Wang G, Yan Y (2012) Improving stability and activity of cross-linked enzyme aggregates based on polyethyleneimine in hydrolysis of fish oil for enrichment of polyunsaturated fatty acids. Appl Biochem Biotechnol 166:925–932

28. Cui JD, Zhang S, Sun LM (2012) Cross-linked enzyme aggregates of phenylalanine ammonia lyase: novel biocatalysts for synthesis of L-phenylalanine. Appl Biochem Biotechnol 167:835–844

29. Wang M, Jia C, Qi W, Yu Q, Peng X, Su R, He Z (2011) Porous CLEAs of papain: application to enzymatic hydrolysis of macromolecules. Bioresour Technol 102:3541–3545

30. Hormigo D, García-Hidalgo J, Acebal C, de la Mata I, Arroyo M (2012) Preparation and characterization of cross-linked enzyme aggregates (CLEAs) of recombinant poly-3-hydroxybutyrate depolymerase from *Streptomyces exfoliatus*. Bioresour Technol 115:177–182

31. Majumder AB, Gupta MN (2010) Stabilization of *Candida rugosa* lipase during transacetylation with vinyl acetate. Bioresour Technol 101:2877–2879

32. Guauque TMP, Foresti ML, Ferreira ML (2013) Cross-linked enzyme aggregates (CLEAs) of selected lipases: a procedure for the proper calculation of their recovered activity. AMB Express 3:25

33. Li L, Li G, Cao LC, Ren GH, Kong W, Wang SD, Guo GS, Liu YH (2015) Characterization of cross-linked enzyme aggregates of a novel beta galactosidase, a potential catalyst for the synthesis of galacto-oligosaccharides. J Agric Food Chem 63:894–901

34. Wilson L, Illanes A, Abian O, Pessela BCC, Fernandez-Lafuente R, Guisán JM (2004) Co-aggregation of penicillin G acylase and polyionic polymers: an easy methodology to prepare enzyme biocatalysts stable in organic media. Biomacromolecules 5:852–857

35. Kim MI, Kim J, Lee J, Jia H, Na HB, Youn JK, Kwak JH, Dohnalkova A, Grate JW, Wang P, Hyeon T, Park HG, Chang HN (2007) Cross-linked enzyme aggregates in hierarchically-ordered mesoporous silica: a simple and effective method for enzyme stabilization. Biotech Bioeng 96:210–218

36. Hilal N, Nigmatullin R, Alpatova A (2004) Immobilization of cross-linked lipase aggregates within microporous polymeric membranes. J Memb Sci 238:131–141

37. Mukherjee J, Gupta MN (2015) Paradigm shifts in our view on inclusion bodies. Curr Biochem Eng 2:1–9

38. Mateo C, Palomo JM, van Langen LM, Rantwijik FV, Sheldon RA (2004) A new, mild cross-linking methodology to prepare cross-linked enzyme aggregates. Biotechnol Bioeng 86:273–276

39. Bell G, Halling PJ, Moore BD, Partridge J, Rees DG (1995) Biocatalyst behavior in low-water systems. Trends Biotechnol 13:468–473

40. Tyagi R, Batra R, Gupta MN (1999) Amorphous enzyme aggregates: stability towards heat and aqueous-organic cosolvent mixtures. Enzyme Microb Technol 24:348–353

41. López-Gallego F, Betancor L, Hidalgo A, Alonso N, Fernandez-Láfuente R, Guisán JM (2005) Co-aggregation of enzymes and polyethyleneimine: a simple method to prepare stable and immobilized derivatives of glutaryl acylase. Biomacromolecules 6:1639–1842

42. Vaidya A, Fischer L (2006) Stabilization of new imprint property of glucose oxidase in pure aqueous medium by cross-linked-imprinting approach. In: Guisan JM (ed) Immobilization of enzymes and cells. Humana Press, NJ, pp 175–183

43. Dalal S, Kapoor M, Gupta MN (2007) Preparation and characterization of combi-CLEAs catalyzing multiple non-cascade reactions. J Mol Catal B Enzymatic 44:128–132

44. Arsenault A, Cabana H, Peter Jones J (2011) Laccase-based CLEAs: Chitosan as a novel cross-linking agent. Enzym Res 2011:376015

45. Fairhead M, Thony-Meyer L (2010) Cross-linking and immobilization of different proteins with recombinant *Verrucomicrobium spinosum* tyrosinase. J Biotechnol 150:546–551

46. Garcia-Garcia MI, Sola-Carvajal A, Sanchez-Carron G, Carcia-Carmona F, Sanchez-Ferrer A (2011) New stabilized FastPrep-CLEAs for sialic acid synthesis. Bioresour Technol 102:6186–6191

47. Chen J, Zhang J, Han B, Li Z, Li J, Feng X (2006) Synthesis of crosslinked enzyme aggregates (CLEAs) in CO_2 expanded micellar solutions. Colloids Surf B Biointerfaces 48:72–76

48. Cui JD, Cui LL, Zhang SP, Zhang YF, Su ZG, Ma GH (2014) Hybrid magnetic cross-linked enzyme aggregates of phenylalanine ammonia lyase from *Rhodotorula glutinis*. PLoS One 9:e97221

49. Cui JD, Li LL, Bian HJ (2013) Immobilization of cross-linked phenylalanine ammonia lyase aggregates in microporous silica gel. PLoS One 8:e80581

50. Talekar S, Ghodake V, Ghotage T, Rathod P, Deshmukh P, Nadar S, Mulla M, Ladole M (2012) Novel magnetic cross-linked enzyme aggregates (magnetic CLEAs) of alpha amylase. Bioresour Technol 123:542–547

51. Jiang Y, Shi L, Huang Y, Gao J, Zhang X, Zhou L (2014) Preparation of robust biocatalyst based on cross-linked enzyme aggregates entrapped in three-dimensionally ordered macroporous silica. ACS Appl Mater Interfaces 6:2622–2628

52. Ning C, Su E, Tian Y, Wei D (2014) Combined cross-linked enzyme aggregates (combi-CLEAs) for efficient integration of a ketoreductase and a cofactor regeneration system. J Biotechnol 184:7–10

53. Jung DH, Jung JH, Seo DH, Ha SJ, Kweon DK, Park CS (2013) One-pot bioconversion of sucrose to trehalose using enzymatic sequential reactions in combined cross-linked enzyme aggregates. Bioresour Technol 130:801–804

54. Ba S, Peter-Jones J, Cabana H (2014) Hybrid bioreactor (HBR) of hollow fibre microfilter membrane and cross-linked laccase aggregates eliminate aromatic pharmaceuticals in waste waters. J Hazard Mater 280:662–670

Chapter 10

Protein-Coated Microcrystals, Combi-Protein-Coated Microcrystals, and Cross-Linked Protein-Coated Microcrystals of Enzymes for Use in Low-Water Media

Joyeeta Mukherjee and Munishwar N. Gupta

Abstract

Protein-coated microcrystals (PCMC) are a high-activity preparation of enzymes for use in low-water media. The protocols for the preparation of PCMCs of Subtilisin Carlsberg and *Candida antarctica* lipase B (CAL B) are described. The combi-PCMC concept is useful both for cascade and non-cascade reactions. It can also be beneficial to combine two different specificities of a lipase when the substrate requires it. Combi-PCMC of CALB and Palatase used for the conversion of coffee oil present in spent coffee grounds to biodiesel is described. Cross-linked protein-coated microcrystals (CL-PCMC) in some cases can give better results than PCMC. Protocols for the CLPCMC of Subtilisin Carlsberg and *Candida antarctica* lipase B (CAL B) are described. A discussion of their applications is also provided.

Key words Enzymes in organic solvents, Enzymes in organic synthesis, Subtilisin Carlsberg, *Candida antarctica* lipase B (CAL B), Monoglyceride synthesis, Biosurfactants, Protein-coated microcrystals, Cross-linked protein-coated microcrystals, Biodiesel from coffee oil, Low-water enzymology

1 Introduction

The successful applications of using enzymes in nearly anhydrous organic solvents have resulted in considerable interest in this technology [1–5]. Several advantages of this approach have been discussed at several places [6–8]. An important outcome is that inexpensive hydrolases (e.g., lipases and proteases) can be used for synthetic purposes. Hence, it is not surprising that applications using hydrolases dominate the use of enzymes in such media. After the initial excitement, an early disappointment was the low catalytic rates which are generally observed in these media. Protein-coated microcrystals (PCMC) were introduced as a high activity preparation [9] and have been a fairly well studied formulation now. It can be viewed as an immobilization method which uses

Shelley D. Minteer (ed.), *Enzyme Stabilization and Immobilization: Methods and Protocols*, Methods in Molecular Biology, vol. 1504, DOI 10.1007/978-1-4939-6499-4_10, © Springer Science+Business Media New York 2017

small microcrystals of low-molecular weight compounds such as sugars, salts, amino acids as core materials. Some examples of such core materials used in the PCMCs of enzymes as well as their applications are given in Table 1.

One essential requirement for using enzymes in such low-water media is that enzyme preparations need to be free from bulk water. Lyophilization or freeze drying has often been used both for obtaining free enzyme or "immobilized enzymes" in "dry" form. Most of the laboratory scale freeze driers tend to be simple in design and hence such preparations from different laboratories may give somewhat different results [10]. Precipitation of enzymes from their aqueous solutions, on the other hand, is easy to reproduce and in fact has emerged as a better option for "drying" enzymes as it quite often results in a more active enzyme preparation [11–14].

PCMCs are made by co-precipitating a mixture of the core material and enzyme together. The core material precipitates faster and forms microcrystals. The protein precipitation follows and the protein is deposited or "coated" over the core microcrystals. It is likely that dispersion of the enzyme over a large microcrystalline surface also is responsible for the high catalytic activity [12].

Co-immobilization of enzymes is a well-known approach to catalyze a cascade of reactions [15, 16]. In the combi-PCMC format, it can be exploited for benefitting from the synergy of two different substrate specificities of the same class of enzyme such as lipases [17]. PCMCs prepared by using two different enzyme solutions together are known as Combi-PCMCs. This gives us the advantage of incorporating the properties of both the enzymes into one formulation and hence makes the enzyme preparation more versatile. For example, *Candida antarctica* lipase B (CAL B) is a relatively expensive enzyme, whereas Palatase (*Rhizomucor miehei* lipase) is an inexpensive one. Combining the two for biodiesel synthesis from coffee oil by transesterification with ethanol gives us cost effectiveness. Palatase alone cannot be used, because it is a 1,3-specific enzyme, so the yield of biodiesel will be less as it will generate 2-monoglyceride in addition to synthesizing free fatty acid ester (biodiesel) by attacking the 1 and 3 position only of the triglyceride (coffee oil). A nonspecific lipase like CAL B will convert this 2-monoglyceride further to give the free fatty acid esters by transesterification. Hence a combi-PCMC accounts for the synergistic effect of both lipases and gives the best results possible [17]. It was also found that the PCMCs of the two enzymes prepared separately and physically mixed before use in the reaction did not give biodiesel yields as good as the combi-PCMCs [17].

Cross-linking an already immobilized enzyme has often resulted in a more stable biocatalyst [18, 19]. It was found that cross-linked protein-coated microcrystals (CLPCMC) in some cases have given better results than even PCMC [20–22]. Like PCMC, CLPCMC technology is also patent protected [22].

Table 1
Core materials used for the preparation of PCMC and CL-PCMC

Sl. no.	Enzyme	Core material	Application	Reference
1.	Subtilisin Carlsberg	K_2SO_4	The catalytic rate for PCMC preparations was over three orders of magnitude higher than typically found using a lyophilized preparation	[9, 12]
2.	*Pseudomonas cepacia* lipase (PS)	K_2SO_4	PCMC-Pseudomonas sp. lipase was found to be 200-fold more active than the straight from the bottle preparation	[9, 28]
3.	*Candida antarctica* lipase B (CAL B) *Mucor miehei* lipase	K_2SO_4	Coated micro-crystals of CALB and lipase from *M. miehei* showed a 5 and 15-fold increase in catalytic activity, respectively as compared to the straight from the bottle enzyme	[9]
4.	*Candida antarctica* lipase A (CAL A)	K_2SO_4	A 60-fold increase in the activity of K_2SO_4– CALA was observed	[9]
5.	Subtilisin Carlsberg (SC)	Solid state buffers	SC-PCMC with organic buffer carriers (Na-AMPSO) showed a threefold greater activity than that observed when using the unbuffered system (PCMC-SC/K_2SO_4). In comparison with freeze-dried preparations, this represents an 3000-fold increase in catalytic activity	[29]
6.	*Pseudomonas cepacia* lipase (PS)	K_2SO_4, KCl, KNO_3, Na_2SO_4, NaCl, $NaNO_3$	PS-PCMC with K_2SO_4 and $NaNO_3$ as core gave a twofold increase in the biodiesel synthesis as compared to free PS	[30]
7.	*Mucor javanicus* lipase (MJ)	K_2SO_4 KCl	MJ-PCMC showed a 26-fold (KCl) and 22-fold (K_2SO_4) increase in the rates of esterification of lauric acid with 1-propanol in isooctane. The precipitating solvent was acetonitrile	[31]
8.	Vaccine diphtheria toxoid, DT	L-glutamine	DT-coated microcrystals showed much higher thermal stability	[32]
9.	Alpha chymotrypsin (AC)	Trehalose	AC-PCMC showed a 9-times increase in transesterification rates in octane as compared to the simple enzyme precipitates in propanol	[12]
10.	*Thermomyces lanuginosus* lipase (TLL)	K_2SO_4 Glucose MOPS Glycine	TLL-CL-PCMC: the highest reactivity of lipase biocatalyst was obtained using glycine as the matrix component with acetone as the precipitating organic solvent for the esterification of palmitic acid and ethanol	[21]

Both PCMC and CLPCMC are useful in low-water media. If the water content of the media becomes high enough for water to appear as a separate phase, it will slowly disintegrate the core and dissolve these preparations.

This chapter describes:

1. The specific protocols for the preparations of PCMCs of a lipase (*Candida antarctica* lipase B, CAL B) and a protease (Subtilisin Carlsberg) [Protocol 3.1]. The lipase was used for the esterification of palmitic acid and glycerol to give monoglyceride with a fairly good % of conversion [23] (Fig. 1). PCMC of Subtilisin Carlsberg was used for the transesterification of *N*-acetyl-L-phenylalanine ethyl ester and *n*-propanol in various solvents [20].

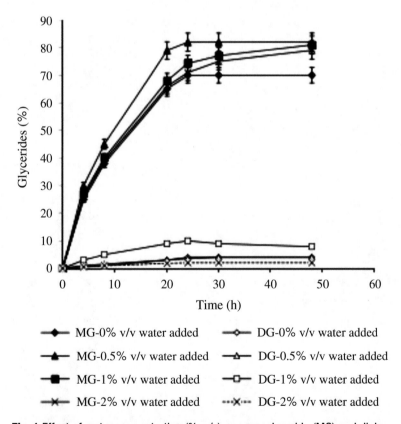

Fig. 1 Effect of water concentration (%, v/v) on monoglyceride (MG) and diglyceride (DG) using PCMCs of *Candida antarctica* lipase B (CALB) in the presence of 20 mg molecular sieves. Experiments were carried out in triplicates and error bars represent the percentage error in each set of readings [Reproduced from ref. [23] with permission from Elsevier]

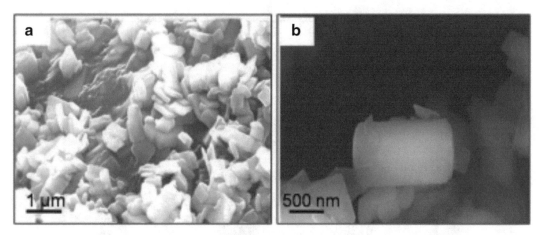

Fig. 2 SEM images showing (**a**) the K$_2$SO$_4$ crystals and (**b**) the Combi-PCMCs of Palatase and CALB prepared using K$_2$SO$_4$ as the core material (Reproduced from ref. [17] with permission from Chemistry Central)

2. Preparation of combi-PCMC of two lipases, CAL B and Palatase (*Rhizomucor miehei* lipase), for converting coffee oil present in spent coffee grounds to biodiesel [17] [Protocol 3.2]. SEM images of the bare K$_2$SO$_4$ crystals and the protein-coated K$_2$SO$_4$ crystals are shown in Fig. 2. Combi-PCMCs were able to give ~85 % biodiesel in 72 h (Fig. 3).

3. CLPCMC of Subtilisin Carlsberg used for the transesterification of *N*-acetyl-L-phenylalanine ethyl ester and *n*-propanol in various solvents [20] and the CLPCMC of CAL B used for monoglyceride synthesis [23] [Protocol 3.3].

2 Materials

1. 3-Å molecular sieves (E. Merck, Mumbai, India).

2. Glutaraldehyde solution: 25 %, v/v in water.

3. Tris–HCl buffer: 50 mM and pH 7.8.

4. Sodium phosphate buffer: 10 mM and pH 7.

5. Subtilisin Carlsberg type VIII (Sigma Aldrich, St. Louis, MO, USA), Cat. No. P5380-100MG.

6. Lipozyme CALB (a commercial liquid preparation of the free form of *Candida antarctica* lipase B) (Novozymes, Denmark).

7. Palatase (a commercial liquid preparation of the free form of *Rhizomucor miehei* lipase B) (Novozymes, Denmark).

8. Dimethoxyethane (Sigma-Aldrich, St. Louis, MO, USA).

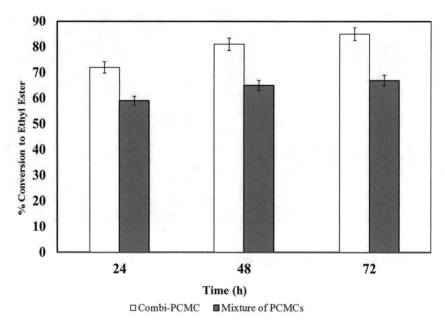

Fig. 3 Biodiesel from clean coffee oil catalyzed by Combi-PCMC (*white bars*) and a mixture of PCMCs (1:1) (*grey bars*) of CALB and Palatase (*tert*-butanol used for precipitation). Reactions were performed taking 0.5 g coffee oil. Ethanol was added in molar ratio 4:1 (ethanol–oil). Conversion rates were measured by analyzing the aliquots withdrawn at the different time intervals using Gas Chromatographic analysis. The experiments were done in duplicates and the error between the two sets was within 3 % (Reproduced from ref. [17] with permission from Chemistry Central)

3   Methods

3.1 Preparation of PCMCs

3.1.1 Preparation of PCMCs of Candida antarctica Lipase B (CAL B)

1. CAL B (0.05 mL) is diluted with 10 mM sodium phosphate buffer (pH 7) (0.05 mL) (*see* **Notes 1–3**).

2. The above enzyme solution is mixed with 0.3 mL saturated K_2SO_4 solution (*see* **Note 4**).

3. This mixture is now added dropwise to ice-chilled organic solvent (*see* **Notes 5** and **6**) under conditions of orbital shaking at 150 rpm and at 4 °C. In this case 1,2-dimethoxyethane (DME) (4 mL) (*see* **Note 7**).

4. The precipitation is allowed to continue for 30 min, the PCMCs so formed are recovered by centrifugation at $9000 \times g$ for 5 min (*see* **Note 8**).

5. The PCMCs are washed thrice with ice-chilled DME (1 mL each time) and finally once with the ice-chilled anhydrous reaction medium (*see* **Notes 9–11**).

3.1.2 Preparation
of PCMCs of Subtilisin
Carlsberg (SC)

1. SC (4 mg) is dissolved in 50 mM Tris–HCl buffer (pH 7.8) (0.1 mL).

2. This is added to 0.3 mL of saturated K_2SO_4 solution.

3. The above mixture is then precipitated dropwise into chilled *n*-propanol (6 mL) under conditions of orbital shaking at 150 rpm and 4 °C.

4. After 30 min, the precipitate is centrifuged at $5000 \times g$ for 5 min.

5. The PCMCs so obtained are washed with dry ice-chilled *n*-propanol or ice-chilled *n*-propanol containing various amounts of water.

6. Finally the PCMCs are washed with the reaction solvent.

3.2 Preparation of Combi-PCMCs (Fig. 2)

The basic protocol for preparation of Combi-PCMC is the same as described in Subheading 3.1. The slight variations are described in detail below.

1. Palatase (0.05 mL) and CALB (0.05 mL) are mixed together and then diluted in 10 mM sodium phosphate buffer (pH 7) (0.1 mL).

2. This enzyme is then mixed with 0.6 mL of saturated K_2SO_4 solution.

3. The resulting solution is added dropwise to *tert*-butanol (6 mL) with constant shaking at 200 rpm at 4 °C (*see* **Notes 12** and **13**).

4. After 30 min, the precipitates are centrifuged at $8000 \times g$ for 5 min.

5. The Combi-PCMCs so obtained are then washed twice with *tert*-butanol and then with the reaction medium (*see* **Note 14**) before use.

3.3 Preparation of CLPCMCs

3.3.1 Preparation
of CL-PCMC of Subtilisin
Carlsberg (SC)

1. PCMC of SC is prepared as described in Subheading 3.1.2 (until **step 5**).

2. The PCMC obtained is dispersed in chilled *n*-propanol (0.5 mL) with mild vortexing.

3. Glutaraldehyde (25 %, v/v; 0.01 mL) was slowly added to the above suspension so that the final concentration of glutaraldehyde in the solution becomes 50 mM.

4. The mixtures are kept at 4 °C for 1 h with continuous orbital shaking at 300 rpm after which they are centrifuged at $5000 \times g$ for 5 min.

5. The CLPCMCs so obtained are washed thrice with dry chilled *n*-propanol or chilled *n*-propanol containing various amounts of water (*see* **Note 15**).

6. Finally the CLPCMCs are washed with the reaction solvent before use.

3.3.2 Preparation of CLPCMC of CAL B

1. PCMC of CAL B is prepared as described in Subheading 3.1.1 (*see* **Note 16**).

2. The PCMCs are then suspended in DME (0.5 mL) by mild vortexing.

3. Glutaraldehyde (25 %, v/v) is slowly added to the above suspension (*see* **Notes 17** and **18**).

4. The mixtures are kept at 4 °C for 1 h with continuous orbital shaking at 300 rpm (*see* **Note 19**).

5. The CLPCMCs are recovered by centrifugation at 4 °C and $8000 \times g$ for 5 min.

6. They are washed thrice with chilled DME (1 mL each time) to ensure the complete removal of the unreacted glutaraldehyde (*see* **Note 20**).

7. Finally, the CLPCMCs are washed with the chilled anhydrous reaction medium or with the reaction medium containing different amounts of added water.

4 Notes

1. For all buffer solutions, membrane filtered water obtained from the membrane filtration unit Labconco WaterPro (LabConco Corporation, Kansas City, Missouri) was used with resistivity of 18.2 MΩ cm and total organic content of less than five parts per billion.

2. The pH of the buffer used should be the optimum pH of the enzyme.

3. In case of an enzyme in the solid form, after dissolving the enzyme in the buffer, it should be centrifuged to remove the insoluble impurities and the clear enzyme solution should be used further. In case of a liquid enzyme, it should be ensured that no precipitate appears on adding the enzyme to the buffer solution. This sometimes occurs when some stabilizers used in the liquid enzyme formulation interact with the buffer ions and precipitate.

4. K_2SO_4 is generally used as a carrier for the PCMCs, because it adsorbs very little water, is cheap and readily available and also it produces fine particles on precipitation and can be handled very easily. The other cores which are used are generally sugars or solid state buffers (Table 1).

5. The mode of addition of the aqueous and the organic phase may be important. The aqueous enzyme solution should be added to the organic precipitant (normal mode) and not the other way round (reverse mode). The two modes of mixing during the precipitation step may result in enzyme preparations which give different transesterification activities in anhydrous

media [14]. The enzymes prepared by the reverse mode of precipitation gave lower rates in case of alpha chymotrypsin and subtilisin [14].

6. Choice of the organic solvent to be used as a precipitant is different for different enzymes. Hence, an initial screen of the precipitants should be carried out. The organic solvent retains all the activity in the precipitate should be chosen. Sometimes this step also leads to the purification of the enzyme as the precipitant may precipitate the enzyme selectively. Hence the specific activity of the preparation may increase many folds.

7. In this case DME was used as the precipitant because screening of the organic solvents showed that DME when used as the precipitant recovers nearly 100% of the activity of CAL B in the precipitate and was better than all other solvents screened [23].

8. It has been found that shaking for 30 min is generally sufficient to precipitate all the protein from the aqueous solution. However, this time of precipitation may need to be optimized for novel proteins.

9. Precipitation in organic solvent removes the bulk water and hence "dries" the enzyme. But the washing step is equally important as it removes the residual water from the enzyme and thus makes the enzyme "dry."

10. Enzymes require some minimum amount of water for their optimum functioning. This can be taken care of either by introducing some minimum added water in the reaction medium or by equilibrating the reaction medium at the optimum a_w [24]. Ideally, during the final wash with the reaction solvent, the solvent should either be equilibrated at the optimum a_w or contain the optimum % of added water.

11. The washing with the reaction medium should be with anhydrous solvent only if no added water is used in the reaction. Kapoor and Gupta [23] used this PCMC of CAL B for the synthesis of monoglycerides in acetone. The reaction was carried out by adding different amounts of water (Fig. 1) in acetone. Hence in each case, the final washing should be with ice-chilled acetone containing x% water (where $x = 0$, 0.5, 1.2 mL) (Fig. 1).

12. In case of Combi-PCMCs, the precipitating organic solvent should be such that it precipitates the maximum activity for both the lipases cumulatively. DME and *tert*-butanol were comparable in precipitating the activity of CAL B but *tert*-butanol was a much better precipitant for Palatase [25]. Hence *tert*-butanol was chosen as the precipitating organic solvent for the preparation of the combi-PCMC.

13. *Tert*-butanol has a freezing point of 25–26 °C. So, initially it was taken at ~27 °C in the form of liquid. Once the mixture of the enzyme and the salt solution was added, the tube could be transferred to an orbital shaker at 200 rpm and at 4 °C for 30 min so that loss of enzyme activity due to temperature does not occur.

14. Sometimes, lipase catalyzed reactions are carried out in solvent free media [26, 27]. In such cases no reaction medium is used. Hence the final wash may be with the precipitating solvent itself. However, if the precipitating solvent has the possibility of acting as a competing substrate in the reaction mixture, then the final wash may be given with a neutral anhydrous chilled organic solvent like acetone. Banerjee et al. [17] used combi-PCMC of CAL B and Palatase for the synthesis of biodiesel from coffee oil and ethanol in solvent free media. In this case, the final wash was with acetone. Sometimes in a solvent free reaction system, one of the substrates, like ethanol in this case, may also be used to give the final wash but there are two concerns: (a) Ethanol generally acts as a denaturant for enzymes (b) the ratio of the substrates, ethanol–oil, may be critical for the optimum reaction conditions and hence any residual ethanol from the final wash may affect the ratio detrimentally. So acetone was the best choice. This preparation gave nearly 90 % conversion to biodiesel from coffee oil (Fig. 3).

15. It is worth noting that for PCMC of SC, 0.5 % (v/v) water in the reaction medium gives the best conversions for the transesterification reaction whereas for CLPCMC, 0 % water works best in *tert*-amyl alcohol (Fig. 4) [20].

16. In Subheading 3.1.1, **step 5** also consists of a final wash with the reaction medium. However, when we are preparing PCMC for cross-linking to obtain CLPCMC, this wash with the reaction medium is not required at this stage.

17. Glutaraldehyde should be stored between 2 and 8 °C. When handling glutaraldehyde-based solutions, avoid contact with the liquid and inhalation of the vapor. Protective gloves, splash-proof monogoggles, and protective clothing should be worn. In most of the cases glutaraldehyde has been used for cross-linking but similar precautions may have to be taken while using other cross-linking agents.

18. The amount of glutaraldehyde (25 %, v/v) used should be optimized so that the final concentration in the precipitating solvent ranges from 50 to 200 mM. The extent of cross-linking affects the performance of the enzyme preparation significantly (Table 2) [23].

19. The cross-linking time may vary from 10 min to 5 h depending on the enzyme to be cross-linked. Thus, the cross-linking time

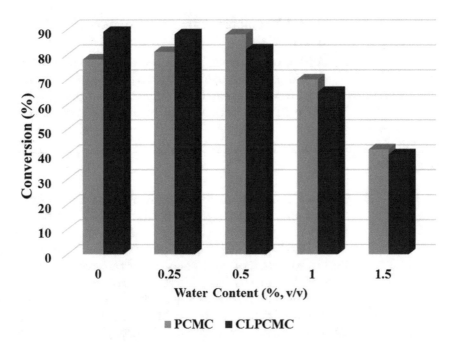

Fig. 4 Effect of water content on transesterification reactions catalyzed by PCMCs and CLPCMCs in tert-amyl alcohol at 70 °C. *N*-acetyl-L-phenylalanine ethyl ester (20 mM) with n-propanol (0.85 M) was taken in 5 mL of *tert*-amyl alcohol (containing different amount of water) followed by addition of different enzyme preparations (each containing 4 mg of SC). The reaction mixture was incubated at 70 °C with constant shaking at 200 rpm and the progress of reaction was monitored after 6 h withdrawing aliquots, which were analyzed by HPLC (Adapted from ref. [20])

Table 2
Effect of CLPCMCs of CALB prepared using different concentrations of glutaraldehyde on monoglyceride synthesis

| Water conc. (%, v/v) | Glutaraldehyde (conc) | | | | | |
| | 50 mM | | 100 mM | | 200 mM | |
	MG (%)[a]	DG (%)	MG (%)[a]	DG (%)	MG (%)[a]	DG (%)
0	47	0.7	50	6.0	58	6.8
0.5	49	4.5	68	4.1	82	5.0
1	73	4.5	84	4.0	87	3.3
2	64	2.5	79	2.3	72	10.0

Esterification of palmitic acid (0.46 g) and glycerol (0.65 g) was carried out in acetone (4 ml) at 50 °C in the presence of 20 mg molecular sieves at 300 rpm in an orbital shaker. Experiments were carried out in triplicate and percentage error in each reading in a set was within 5 %. (Reproduced from ref. [23] with permission from Elsevier)
[a]Conversions were determined using GC after 24 h. No triglyceride formation could be detected even after 24 h

should be optimized for the preparation of CLPCMCs in the individual cases.

20. After the cross-linking reaction is over, the enzyme formulations should be washed thoroughly with buffer to remove the unreacted cross-linking reagent. If this step is not done properly, it may interfere with the activity assays of the cross-linked preparations.

Acknowledgments

This work was supported by funds obtained from Department of Science and Technology [Grant No.: SR/SO/BB-68/2010] and Department of Biotechnology [Grant No.: BT/PR14103/BRB/10/808/2010], both Government of India organizations. Finally, we thank past members of our research group; Dr. Aparna Sharma, Dr. Shweta Shah, Dr. Kusum Solanki, Dr. Manali Kapoor, Veena Singh, and Aditi Banerjee, whose data have been described/quoted in this chapter. We would also like to acknowledge the collaboration in this area with Prof. Peter J. Halling (University of Strathclyde, Glasgow, UK) funded by UKIERI, UK (2008–2012).

References

1. Klibanov AM (2001) Improving enzymes by using them in organic solvents. Nature 409:241–246

2. Mattiasson B, Adlercreutz P (1991) Tailoring the microenvironment of enzymes in water-poor systems. Trends Biotechnol 9:394–398

3. Gupta MN (ed) (2000) Non-aqueous enzymology. Springer, Heidelberg

4. Carrea G, Riva S (2000) Properties and synthetic applications of enzymes in organic solvents. Angew Chem Int Ed 39:2226–2254

5. Carrea G, Riva S (2008) Medium engineering. In: Gotor V, Alfonso I, Garcia-Urdiales E (eds) Asymmetric organic synthesis with enzymes. Wiley-VCH, Weinheim, pp 3–20

6. Zaks A, Klibanov AM (1985) Enzyme-catalyzed processes in organic solvents. Proc Natl Acad Sci U S A 82:3192–3196

7. Gupta MN (1992) Enzyme function in organic solvents. Eur J Biochem 203:25–32

8. Dordick JS (1992) Designing enzymes for use in organic solvents. Biotechnol Prog 8:259–267

9. Kreiner M, Moore BD, Parkar MC (2001) Enzyme-coated micro-crystals: a 1-step method for high activity biocatalyst preparation. Chem Commun 12:1096–1097

10. Mukherjee J, Mishra P, Gupta MN (2015) Urea treated subtilisin as a biocatalyst for trans-

formations in organic solvents. Tetrahedron Lett 56:1976–1981

11. Roy I, Gupta MN (2004) Preparation of highly active alpha-chymotrypsin for catalysis in organic media. Bioorg Med Chem Lett 14:2191–2193

12. Solanki K, Gupta MN (2008) Optimizing biocatalyst design for obtaining high transesterification activity by a-chymotrypsin in non-aqueous media. Chem Cent J 2:1–7

13. Majumder AB, Gupta MN (2011) Increasing catalytic efficiency of *Candida rugosa* lipase for the synthesis of tert-alkyl butyrates in low water media. Biocatal Biotrasform 29:238–245

14. Solanki K, Gupta MN, Halling PJ (2012) Examining structure-activity correlations of some high activity enzyme preparations for low water media. Bioresour Technol 115:147–151

15. Roy I, Gupta MN (2004) Hydrolysis of starch by a mixture of glucoamylase and pullulanase entrapped individually in calcium alginate beads. Enzym Microb Technol 34:26–32

16. Rocha-Martín J, Rivas B, Muñoz R, Guisán JM, López-Gallego F (2012) Rational co-immobilization of bi-enzyme cascades on porous supports and their applications in bio-redox reactions with in situ recycling of soluble cofactors. ChemCatChem 4:1279–1288

17. Banerjee A, Singh V, Solanki K, Mukherjee J, Gupta MN (2013) Combi-protein coated

microcrystals of lipases for production of biodiesel from oil from spent coffee grounds. Sust Chem Processes 1:1–14

18. Fernandez-Lorente G, Fernandez-Lafuente R, Armisen P, Sabuquillo P, Mateo C, Guisan JM (2000) Engineering of enzymes via immobilization and post immobilization techniques: preparation of enzyme derivatives with improved stability in organic media. In: Gupta MN (ed) Methods in non-aqueous enzymology. Birkhauser, Basel

19. Mateo C, Grazu V, Palomo JM, Lopez-Gallego F, Fernandez-Lafuente R, Guisan JM (2007) Immobilization of enzymes on heterofunctional epoxy supports. Nat Protoc 2:1022–1033

20. Shah S, Sharma A, Gupta MN (2008) Cross-linked protein-coated microcrystals as biocatalysts in non-aqueous solvents. Biocatal Biotransform 26:266–271

21. Raita M, Laothanachareon T, Champreda V, Laosiripojana N (2011) Biocatalytic esterification of palm oil fatty acids for biodiesel production using glycine-based cross-linked protein coated microcrystalline lipase. J Mol Catal B Enzym 73:74–79

22. Gupta MN, Shah S, Sharma A (2006) A method for preparation of cross-linked protein coated microcrystals (Appln. No. 2046/DEL/2006 dated 18-9-2006)

23. Kapoor M, Gupta MN (2012) Obtaining monoglycerides by esterification of glycerol with palmitic acid using some high activity preparations of Candida antarctica lipase B. Process Biochem 47:503–508

24. Halling PJ (1992) Salt hydrates for water activity control with biocatalysts in organic media. Biotechnol Tech 6:271–276

25. Schoevaart R, Wolbers MW, Golubovic M, Ottens M, Kieboom AP, van Rantwijk F, van der Wielen LA, Sheldon RA (2004) Preparation, optimization, and structures of cross-linked enzyme aggregates (CLEAs). Biotechnol Bioeng 87:754–762

26. Kumari V, Shah S, Gupta MN (2007) Preparation of biodiesel by lipase-catalyzed transesterification of high free fatty acid containing oil from madhuca indica. Energ Fuels 21:368–372

27. Chiaradia V, Paroul N, Cansian RL, Júnior CV, Detofol MR, Lerin LA, Oliveira JV, Oliveira D (2012) Synthesis of eugenol esters by lipase-catalyzed reaction in solvent-free system. Appl Biochem Biotechnol 168:742–751

28. Solanki K, Gupta MN (2011) A chemically modified lipase preparation for catalyzing the transesterification reaction in even highly polar organic solvents. Bioorg Med Chem Lett 21:2934–2936

29. Kreiner M, Parker MC (2004) High-activity biocatalysts in organic media: solid-state buffers as the immobilisation matrix for protein-coated microcrystals. Biotechnol Bioeng 87:24–33

30. Zheng J, Xu L, Liu Y, Zhang X, Yan Y (2012) Lipase-coated K_2SO_4 micro-crystals: preparation, characterization, and application in biodiesel production using various oil feedstocks. Bioresour Technol 110:224–231

31. Wu JC, Yang JX, Zhang SH, Chow Y, Talukder MR, Choi WJ (2009) Activity, stability and enantioselectivity of lipase-coated microcrystals of inorganic salts in organic solvents. Biocatal Biotransform 27:283–289

32. Murdan S, Somavarapu S, Ross AC, Alpar HO, Parker MC (2005) Immobilisation of vaccines onto micro-crystals for enhanced thermal stability. Int J Pharm 296:117–121

Chapter 11

Macroporous Poly(GMA-co-EGDMA) for Enzyme Stabilization

Nenad B. Milosavić, Radivoje M. Prodanović, Dušan Velićković, and Aleksandra Dimitrijević

Abstract

One of the most used procedures for enzyme stabilization is immobilization. Although immobilization on solid supports has been pursued since the 1950s, there are no general rules for selecting the best support for a giving application. A macroporous copolymer of ethylene glycol dimethacrylate and glycidyl methacrylate (poly (GMA-co-EGDMA)) is a carrier consisting of macroporous beads for immobilizing enzymes of industrial potential for the production of fine chemicals and pharmaceuticals.

Key words Poly (GMA-co-EGDMA), Macroporous, Enzyme stabilization, Covalent immobilization, Glutaraldehyde, Periodate

1 Introduction

Biocatalysts have competed and will compete with conventional chemical catalysts. Advantages of biocatalysts are their high specificity, high activity under mild environmental conditions, and high turnover number. Their biodegradable nature and their label as natural products have become now also very important assets of biocatalysts [1–5]. Drawbacks are inherent to their complex molecular structure that makes them costly to produce and also intrinsically unstable [6]. To improve enzyme stability, several approaches have been developed: the use of stabilizing additives [7], chemical modification [8], selection of enzymes from thermophilic organisms [9], and immobilization [10–12]. Enzyme immobilization is one of the most attractive methods to avoid the problems inherent in the use of free enzymes. Immobilization also facilitates the development of continuous bioprocesses and enhances enzymes properties such as thermostability and activity in nonaqueous media. In particular, the covalent binding of enzyme to solid supports can effectively extend the lifetime of the biocatalysts by protecting the native three-dimensional (3D) structure of enzyme

Shelley D. Minteer (ed.), *Enzyme Stabilization and Immobilization: Methods and Protocols*, Methods in Molecular Biology, vol. 1504, DOI 10.1007/978-1-4939-6499-4_11, © Springer Science+Business Media New York 2017

molecules. The method of immobilization should be gentle, in order not to inactivate the enzyme, and enable binding of enzyme, as much as possible to the support. Epoxy containing supports are almost ideal matrices to perform very simple immobilization of enzymes at both laboratory and industrial scale. A macroporous copolymer of ethylene glycol dimethacrylate and glycidyl methacrylate poly (GMA-co-EGDMA) has been previously used for enzyme stabilization by immobilization [13–15]. It offers a large surface area, hydrophilic microenvironment and number of active groups suitable for immobilization of enzymes. The support does not swell when hydrated, so it can be switched from aqueous media to anhydrous media without significant changes in volume or geometry. Additionally, macroporous poly (GMA-co-EGDMA) internal geometry offers large internal plane surfaces where the enzyme may undergo strong interaction with the support. Finally, it has the possibility of obtaining targeted porous structures [16, 17]. The large number of reactive groups on the carrier could allow multipoint attachment and consequently stabilization of the immobilized enzyme [18]. Enzyme are usually immobilized on poly(GMA-co-EGDMA) through their amino, sulfhydryl, hydroxyl, and phenolic groups.

2 Materials

For this study we used industrial amyloglucosidase (glucoamylase; GA; exo-1,4-α-D-glucosidase; EC 3.2.1.3 from *Aspergillus niger*) 134 IU/mg.

2.1 Enzyme Activity Determination

1. Starch solution: 4% starch (w/v) in 0.05 M sodium acetate buffer, pH 4.5.

2. 3,5-dinitrosalicylic acid (DNS): 5% (w/v).

3. NaOH: 2 M.

4. Sodium potassium tartarate solution: 60%.

5. DNS reagent: Mix 100 mL of 5% (w/v) DNS in 2 M NaOH with 250 mL of 60% (w/v) sodium potassium tartarate and make total volume up to 500 mL with distilled water.

6. Folin–Ciocalteau's phenol reagent and bovine serum albumin (BSA) was used for determination of proteins (*see* **Note 1**).

2.2 Preparation and Activation of Macroporous Poly (GMA-co-EGDMA) Microspheres

1. The monomer phase containing a monomer mixture (GMA and EGDMA) (*see* **Note 1**).

2. Initiator: (azobisisobutyronitrile 1% (w/w)) and inert component (90% (w/w) cyclohexanol and 10% (w/w) tetradecanol) were suspended in 240 g of 1% (w/w) aqueous solution of polyvinylpyrrolidone, PVP.

   3. 1,2, diaminoethane: 1 M in water.

   4. Ethanol.

2.3 Immobilization of Glucoamylase

1. NaCl solution: 1 M in water.

2. Sodium acetate buffer: pH 4.5 and 0.05 M.

3. 2.5 % (w/v) glutaraldehyde in 0.05 M sodium phosphate buffer pH 8.

4. NaCl in buffer: 1 M NaCl in 0.05 M sodium acetate buffer pH 4.5.

5. Glucoamylase: 1 mg/mL in 0.05 M sodium acetate buffer pH 5.0.

6. Sodium periodate solution: 5 mM in 0.05 M sodium acetate buffer pH 5.0.

7. Ethylene glycol: 10 mM.

3 Methods

3.1 Preparation of Macroporous Poly (GMA-co-EGDMA) Microspheres

In this study we use macroporous poly (GMA-co-EGDMA) with pore size of 53 nm and surface area of 50 m^2. The microspheres are prepared by suspension copolymerization in a typical reactor of 0.5 dm^3 in volume.

1. The monomer phase (80.9 g), initiator, and inert component are suspended in aqueous solution of PVP. The ratio of inert component to monomer mixture was 1.32.

2. Polymerization is carried out at 70 °C for 2 h and at 80 °C for 6 h with a stirring rate of 200 rpm (*see* **Note 2**).

3. After completion of the reaction, the copolymer particles are washed with water and ethanol, kept in ethanol for 12 h, and then dried under the vacuum over 45 °C. In this way epoxy group containing copolymer is obtained (*see* **Notes 3** and **4**).

3.2 Activation of Polymer

1. The dry polymer (10 g) is degassed in water for 15 min by a vacuum pump.

2. Fifty milliliters of 1 M 1.2-diaminoethane in water is added to the suction dried polymer.

3. The reaction mixture is heated at 60 °C for 4 h.

4. After heating, the polymer is washed five times with 200 mL of water, three times with the same volume of ethanol and then dried at 45 °C. The concentration of ionizable groups after modification of polymer epoxy group with 1,2-diaminoethane is 1.2 mmol/g (*see* **Note 5**).

3.3 Oxidation of Glucoamylase by Na-Periodate

This method can be applied only if enzyme or protein of interest for immobilization is a glycoenzyme or glycoprotein. In a preliminary study, the effect of oxidant concentration must be determined for each specific glycoenzyme (*see* **Note 6**).

1. Glucoamylase is oxidized by sodium periodate by incubating 10 mg/mL of enzyme with 5 mM $NaIO_4$ in acetate buffer, pH 5.0, for 6 h in the dark at 4 °C. The reaction mixture is stirred occasionally.

2. The reaction is quenched with 10 mM ethylene glycol for 30 min.

3. To remove the by-products, the oxidized glucoamylase solution is dialyzed against 0.05 m sodium acetate buffer, pH 4.5 overnight.

3.4 Reducing Sugar Assay

1. 0.5 mL of DNS reagents (*see* **Note 7**) is added to 0.5 mL of sample with glucose concentration from 0 to 10 mM.

2. The tube is placed in a boiling water bath and the solution heated at 100 °C for 5 min.

3. The tube is rapidly cooled in ice to room temperature.

4. The blank is 0.5 mL of water instead of a solution of glucose, heated as above.

5. Absorbance is read at 540 nm (*see* **Note 8**).

3.5 Determination of Glucoamylase Activity

Activity of soluble glucoamylase was determined according to the Bernefeld method, at 60 °C using 0.05 M sodium acetate buffer, pH 4.5.

1. The immobilized enzyme activity is assayed by incubating an appropriate amount of suction dried derivate ($\approx$50 mg) with 25 mL of 4% (w/v) solution of soluble starch in 0.05 M sodium acetate buffer pH 4.5 with constant stirring.

2. Samples of reaction mixtures (0.5 mL) are withdrawn every 5 min and poured on 0.5 mL of dinidrosalicylic (DNS).

3. The reducing sugars formed this way are then quantified again by the Bernfeld method. One glucoamylase unit is defined as the amount of enzyme which releases reducing carbohydrates equivalent to 1 μmol glucose from soluble starch in 1 min at pH 4.5 and 60 °C.

3.6 Immobilization of Glucoamylase

Three different methods for covalent immobilization of glucoamylase on macroporous poly (GMA-co-EGDMA) are presented on Fig. 1. In this study glucoamylase is used as a model enzyme system, but many other enzymes can be immobilized at the similar way.

3.6.1 Glucoamylase Immobilization via Oxirane Groups (Method I)

This is a conventional method for enzyme immobilization on epoxide containing supports. Poly (GMA-co-EGDMA) binds proteins via its oxirane groups which react at neutral and alkaline pH with the amino groups of the protein molecules, e.g., those of an enzyme to form covalent bonds which are long-term stable within a pH range of 1-12 (Fig. 1 (Method I)). Due to the high density of oxirane groups on the surface of the beads, enzymes are immobilized at various sites of their structure. This phenomenon is called "multi-point-attachment," which is considered a major factor for the high operational stability of enzymes bound to poly (GMA-co-EGDMA) [19].

The method involves the direct enzyme binding on support via oxirane groups.

1. Poly (GMA-co-EGDMA) is incubated with 50 mL of native (unmodified) glucoamylase solution in 1.0 M sodium phosphate buffer (*see* **Note 9**) (1 mg/mL, pH 7.0) at 25 °C.

2. After incubation for 48 h (*see* **Note 10**), the beads are collected by vacuum filtration using glass filter (Whatman).

3. Then, they are washed with 1 M NaCl (3×20 mL) and afterwards with sodium acetate buffer (0.05 M, pH 4.5, 3×20 mL) (*see* **Note 11**).

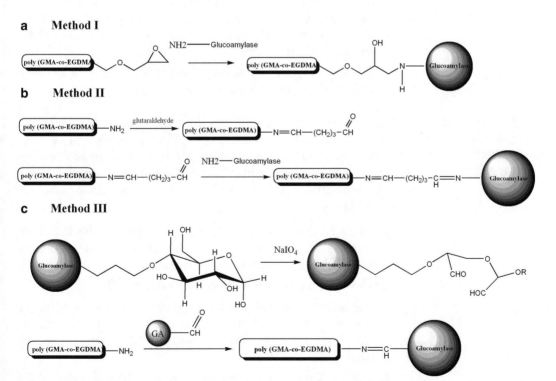

Fig. 1 Schematic illustration of the three different covalent methods of the glucoamylase immobilization of poly (GMA-co-EGDMA) supports

4. The washed beads are combined and collected for determination of enzyme and protein concentrations.

3.6.2 Glucoamylase Immobilization by Glutaraldehyde (Method II)

The immobilization procedure consisted of the three main steps: polymer activation; pretreatment with 1,2-diaminoethane, glutaraldehyde activation of polymer, and enzyme coupling to polymer. As shown in Fig. 1, glutaraldehyde reacts with the amino group of the poly (GMA-co-EGDMA), and the amino group of the enzyme is then coupled to the support via free carbonyl group of the supports bound dialdehyde.

1. Activated acrylic particle (activation with 1,2-diaminoethane) is incubated in 2% (w/v) glutaraldehyde (*see* **Note 12**) in 50 mM sodium phosphate buffer at pH 8 for 2 h at room temperature with occasional stirring.

2. After that incubation, the polymer is washed three times with water and three times with 0.05 M sodium phosphate buffer pH 8.0 (*see* **Notes 13** and **14**).

3. Activated polymers are then incubated with native glucoamylase solution (100 mg of glucoamylase g^{-1} support) in 0.1 M sodium phosphate buffer pH 7.0 for 48 h under slow shaking.

4. After binding, any unbound enzyme is removed by washing the support three times with 100 mL 1 M NaCl, afterwards three times with 100 mL, 0.1 M sodium acetate buffer pH 4.5 and stored in this buffer at 4 °C until used.

5. The washed particles are combined and collected for determination of enzyme and protein concentrations.

3.6.3 Glucoamylase Immobilization by Periodate Method (Method III)

This procedure can be only applied if the enzyme of interest is a glycoenzyme. The immobilization procedure consisted of three main steps: oxidation of glucoamylase by sodium periodate, support treatment with 1,2-diaminoethane, and coupling of oxidized enzyme to poly (GMA-co-EGDMA) with amino group, as shown in Fig. 1c.

1. Oxidized glucoamylase (100 mg) is reacted, in 0.05 M sodium acetate buffer pH 4.5 with 1 g of activated poly (GMA-co-EGDMA), (support with amino group).

2. The reaction mixture is shaken for 48 h at 4 °C.

3. The non-covalently bound enzyme is removed by washing the polymer first with 10 mL of 1 M NaCl solution three times, and then three times with 10 mL of 0.05 M sodium acetate buffer, pH 4.5.

4. The washed polymer particles are combined and collected for determination of enzyme and protein concentrations.

3.7 Activity Coupling Yield

Samples of enzyme solution before and after the immobilization, together with the washings solution, were taken for enzyme assay and protein determination (*see* **Note 15**). The efficiency of immobilization was calculated in terms of enzyme and activity coupling yield. The enzyme coupling yield, η_{enz} (%) and activity coupling yield, η_{act} (%) were calculated as follows:

$$\eta_{enz}(\%) = \frac{P_1}{P_0} \times 100$$

$$\eta_{act}(\%) = \frac{SA_2}{SA_1} \times 100$$

Where P_1 is the amount of immobilized protein (mg), P_0 the initial amount of protein (mg), SA_2 (IU mg^{-1} protein) the specific activity of immobilized glucoamylase, and SA_1 (IU mg^{-1} protein) is specific activity of free lipase (*see* **Note 16**).

4 Notes

1. Other protein assays may be used besides the Lowry assays, e.g., Bradford, BCA, and absorbance at 280 nm.

2. Avoid magnetic stirring of reaction mixture. Other stirring devices should be used (e.g., shaker, orbital stirring, and mechanical stirring)

3. The poly (GMA-co EGDMA) should be kept dry until use; the epoxy groups can hydrolyze.

4. The dry poly (GMA-co EGDMA) will absorb about three times it weight in water (a gram will have a mass of about 4 g when wet).

5. This step is crucial for obtaining active enzyme. It is usually in the range between 1 and 30 µmol of NaIO$_4$ per 1 mg of enzyme.

6. Amino poly (GMA-co-EGDMA) can be stored in a moist form at 4 °C almost unlimited period of time.

7. For obtaining more precise results, DNS reagent is better to be ready prepared.

8. Although the DNS assay is very useful, it does have several disadvantages. It is not accurate for reducing sugar concentrations less than 0.1 g/L (glucose equivalent) and it does not give the same response to every reducing sugar. For example, equivalent molar concentrations of glucose and xylose give significantly different absorbance readings with this assay. Therefore, when mixtures of sugars are present, it may be difficult to interpret the results.

9. The ionic strength and the pH of the enzyme solution can significantly affect of the loading amount and residual activity. Often a high ionic strength (1 M sodium chloride) gives better binding, but this is dependent on the enzyme and the range of ionic strengths and pH values should be evaluated for each enzyme.

10. The binding of the protein to the support can be monitored by performing protein assays on the supernatant and comparing to the initial protein concentration. For some enzymes, incubation times longer than 48 h may be necessary to achieve maximal enzyme binding.

11. Avoid magnetic stirring of polymer. Magnetic stirring acts as a mill stone. Other stirring devices should be used (e.g., shaker, orbital stirring, and mechanical stirring).

12. A higher concentration of GA can be used and this may increase the binding capacity of the support material. However, since GA is often detrimental to enzyme, more extensive washing should be performed.

13. A saturated solution of 2,4 dinitrophenylhydrazine in 0.2 M HCl can be used to detect residual GA in washing solutions. Add 0.2 mL of solution used to wash the GA-activated support to 0.5 mL of saturated 2,4 dinitrophenylhydrazine solution. The formation of yellow precipitate indicates the presence of GA and the support material should be further washed.

14. The GA activated support can be stored in a moist form at 4 °C for at least 1 year without significant lost in binding capacity.

15. If the wash solution contains compounds that interfere with the protein assay, unreliable data may result.

16. For more accurate results, the standard and immobilized enzyme samples should be done in at least triplicate.

Acknowledgements

We wish to thank Professor Slobodan M. Jovanović for development and preparation this new type of support for enzyme immobilization and stabilization as well as Professor Ratko M. Jankov for his helpful comments and information. This work was supported by Ministry of Science of Serbia (projects no. 172049 and no. 46010).

References

1. Klibanov AM (1979) Enzyme stabilization by immobilization. Anal Biochem 93:1–25

2. Velickovic D, Dimitrijevic A, Bihelovic F, Bezbradica D, Jankov R, Milosavic N (2011) A highly efficient diastereoselective synthesis of alpha-isosalicin by maltase from Saccharomyces cerevisiae. Process Biochem 46(8):1698–1702. doi:10.1016/j.procbio.2011.05.007

3. Velickovic D, Dimitrijevic A, Bihelovic F, Bezbradica D, Knezevic-Jugovic Z, Milosavic N (2012) Novel glycoside of vanillyl alcohol, 4-hydroxy-3-methoxybenzyl-alpha-d-glucopyranoside: study of enzymatic synthesis, in vitro digestion and antioxidant activity. Bioproc Biosyst Eng 35(7):1107–1115. doi:10.1007/s00449-012-0695-3

4. Stojanovic M, Velickovic D, Dimitrijevic A, Milosavic N, Knezevic-Jugovic Z, Bezbradica D (2013) Lipase-catalyzed synthesis of ascorbyl oleate in acetone: optimization of reaction conditions and lipase reusability. J Oleo Sci 62(8):591–603. doi:10.5650/Jos.62.591

5. Pavlovic M, Dimitrijevic A, Trbojevic J, Milosavic N, Gavrovic-Jankulovic M, Bezbradica D, Velickovic D (2013) A study of transglucosylation kinetic in an enzymatic synthesis of benzyl alcohol glucoside by alpha-glucosidase from S-cerevisiae. Russ J Phys Chem 87(13):2285–2288. doi:10.1134/S0036024413130207

6. Palestro E (1989) Enzymes in the fine chemical industry: dream and realities. Biotechnology 7:1238–1241. doi:10.1038/nbt1289-1238

7. Murakami Y, Hoshi R, Hirata A (2003) Characterization of polymer-enzyme complex as a novel biocatalyst for nonaqueous enzymology. J Mol Catal B Enzym 22(1–2):79–88. doi:10.1016/S1381-1177(03)00009-2

8. Koops BC, Verheij HM, Slotboom AJ, Egmond MR (1999) Effect of chemical modification on the activity of lipases in organic solvents. Enzyme Microb Tech 25(7):622–631. doi:10.1016/S0141-0229(99)00090-3

9. Cowan DA (1997) Thermophilic proteins: stability and function in aqueous and organic solvents. Comp Biochem Phys A 118(3):429–438. doi:10.1016/S0300-9629(97)00004-2

10. Prodanovic R, Milosavic N, Jovanovic S, Prodanovic O, Velickovic TC, Vujcic Z, Jankov RM (2006) Activity and stability of soluble and immobilized alpha-glucosidase from baker's yeast in cosolvent systems. Biocatal Biotransfor 24(3):195–200. doi:10.1080/10242420600655903

11. Milosavic N, Dimitrijevic A, Velickovic D, Bezbradica D, Knezevic-Jugovic Z, Jankov R (2012) Application of alginates in cell and enzyme immobilization. In: Molina ME, Quiroga AJ (eds) Alginates: production, types and applications. Nova Science Publishers, Hauppauge, NY, pp 37–60

12. Dimitrijevic A, Velickovic D, Rikalovic M, Avramovic N, Milosavic N, Jankov R, Karadzic I (2011) Simultaneous production of exopolysaccharide and lipase from extremophylic Pseudomonas aeruginosa san-ai strain: a novel approach for lipase immobilization and purification. Carbohyd Polym 83(3):1397–1401. doi:10.1016/j.carbpol.2010.10.005

13. Milosavic N, Prodanovic R, Jovanovic S, Novakovic I, Vujic Z (2005) Preparation and characterization of two types of covalently immobilized amyloglucosidase. J Serb Chem Soc 70(5):713–719. doi:10.2298/Jsc0505713m

14. Prodanovic R, Jovanovic S, Vujcic Z (2001) Immobilization of invertase on a new type of macroporous glycidyl methacrylate. Biotechnol Lett 23(14):1171–1174. doi:10.1023/A:1010560911400

15. Prodanovic RM, Milosavic NB, Jovanovic SM, Velickovic TC, Vujcic ZM, Jankov RM (2006) Stabilization of alpha-glucosidase in organic solvents by immobilization on macroporous poly(GMA-co-EGDMA) with different surface characteristics. J Serb Chem Soc 71(4):339–347. doi:10.2298/Jsc0604339p

16. Jovanovic SM, Nastasovic A, Jovanovic NN, Jeremic K (1996) Targeted porous structure of macroporous copolymers based on glycidyl methacrylate. Mater Sci Forum 214:155–162

17. Milosavic N, Pristov JB, Velickovic DV, Dimitrijevic AS, Kalauzi A, Radotic K (2012) Study of the covalently immobilized amyloglucosidase on macroporous polymer by mathematical modeling of the pH optima. J Chem Technol Biotechnol 87(10):1450–1457. doi:10.1002/Jctb.3768

18. Mateo C, Abian O, Fernandez-Lafuente R, Guisan JM (2000) Increase in conformational stability of enzymes immobilized on epoxy-activated supports by favoring additional multipoint covalent attachment. Enzyme Microb Tech 26(7):509–515. doi:10.1016/S0141-0229(99)00188-X

19. Milosavic N, Prodanovic R, Jovanovic S, Vujcic Z (2007) Immobilization of glucoamylase via its carbohydrate moiety on macroporous poly(GMA-co-EGDMA). Enzyme Microb Tech 40(5):1422–1426. doi:10.1016/j.enzmictec.2006.10.018

Chapter 12

Cytochrome *c* Stabilization and Immobilization in Aerogels

Amanda S. Harper-Leatherman, Jean Marie Wallace, and Debra R. Rolison

Abstract

Sol–gel-derived aerogels are three-dimensional, nanoscale materials that combine large surface area with high porosity. These traits make them useful for any rate-critical chemical process, particularly sensing or electrochemical applications, once physical or chemical moieties are incorporated into the gels to add their functionality to the ultraporous scaffold. Incorporating biomolecules into aerogels, other than such rugged species as lipases or cellulose, has been challenging due to the inability of most biomolecules to remain structurally intact within the gels during the necessary supercritical fluid (SCF) processing. However, the heme protein cytochrome *c* (cyt.*c*) forms self-organized superstructures around gold (or silver) nanoparticles in buffer that can be encapsulated into wet gels as the sol undergoes gelation. The guest–host wet gel can then be processed to form composite aerogels in which cyt.*c* retains its characteristic visible absorption. The gold (or silver) nanoparticle-nucleated superstructures protect the majority of the protein from the harsh physicochemical conditions necessary to form an aerogel. The Au~cyt.*c* superstructures exhibit rapid gas-phase recognition of nitric oxide (NO) within the bioaerogel matrix, as facilitated by the high-quality pore structure of the aerogel, while remaining viable for weeks at room temperature. More recently, careful control of synthetic parameters (e.g., buffer concentration, protein concentration, SCF extraction rate) have allowed for the preparation of cyt.*c*–silica aerogels, sans nucleating nanoparticles; these bioaerogels also exhibit rapid gas-phase sensing while retaining protein structural stability.

Key words Cytochrome *c*, Aerogel, Gold nanoparticle, Ultraviolet–visible spectroscopy, TEM, Nitric oxide

1 Introduction

Aerogels are sol–gel-based ultraporous materials, typically metal oxides [1], that take the form of covalently bonded nanoscale solid mesh networks, co-continuous with 3D-plumbed free volumes [2–4]. Drying fluid-filled gels through supercritical solvent extraction creates aerogels, which distinguishes them from densified xerogels that are dried under ambient conditions. The supercritical drying procedure minimizes pore collapse, and the resulting aerogels have 80–99% open porosity with gas-phase analyte response times close to open-medium diffusion rates [5]. These traits make

Shelley D. Minteer (ed.), *Enzyme Stabilization and Immobilization: Methods and Protocols*, Methods in Molecular Biology, vol. 1504, DOI 10.1007/978-1-4939-6499-4_12, © Springer Science+Business Media New York 2017

them excellent sensor platforms, as well as electrodes for battery, supercapacitor, and fuel-cell applications once physical or chemical moieties are incorporated into the pore–solid nanoarchitecture to add energy-storage or -conversion functionality [2, 5–13].

Adding biofunctionality to aerogels is challenging because of the necessary supercritical fluid processing that compromises the structural integrity of most biomolecules encapsulated in silica gels, with notable exceptions being the enzyme, lipase [14], and cellulose [15]. However, the heme protein cyt.c forms self-organized superstructures as nucleated by specific adsorption of the protein at gold (or silver) nanoparticles in buffered medium (abbreviated as Au~cyt.c superstructures in the case of gold nanoparticles, *see* Fig. 1). These metal–protein superstructures can then be encapsulated as guests within silica gels during the sol-to-gel transition, and processed to form aerogels in which cyt.c retains its characteristic visible absorption [16–19]. The Au~cyt.c super-structures even exhibit rapid gas-phase recognition of nitric oxide (NO) within the aerogel matrix, facilitated by the high-quality pore structure of the aerogel [17]. These bioaerogels do not need to be kept wet or stored at 4 °C to maintain viability of the cyto-chrome. The organization of protein within the Au~cyt.c super-structure imparts vital protection against protein denaturation during the harsh physical and chemical processing conditions nec-essary to form the aerogel. Thousands of cyt.c molecules per gold or silver nanoparticle are stabilized within each superstructure.

Initial discovery of these self-assembled protein superstruc-tures arose, when the intensity of the Soret band measured for the Au + cyt.c + silica composite aerogel revealed that most of the cyt.c guests survived the SCF processing. A nanoparticle-nucleated protein superstructure was deposited and then verified by trans-mission electron microscopy. By titrating buffered media contain-ing cyt.c plus Au or Ag nanoparticles with guanidinium hydrochloride, which unfolds the protein, we demonstrated that even in the absence of the aerogel scaffold, the self-assembled pro-tein is stabilized to unfolding relative to unassociated cyt.c in buff-ered solution. These results provide further evidence of the existence of Au~cyt.c superstructures in open medium [17]. Similar protein structures, comprising a protein adsorption layer assembled on the surface of colloidal nanoparticles were later reported and labeled as "protein coronas" [20–22]. These coro-nas consist of two classes of bound proteins. Hard corona proteins are tightly bound proteins that interact directly with nanomaterial surfaces and tend not to desorb while soft corona proteins are proteins that interact with hard corona proteins via weak protein–protein interactions [23].

Recently, refinement of the synthetic parameters, namely buf-fer and protein concentrations, and SCF extraction rate, has per-mitted the preparation of cyt.c–silica aerogels, sans nucleating

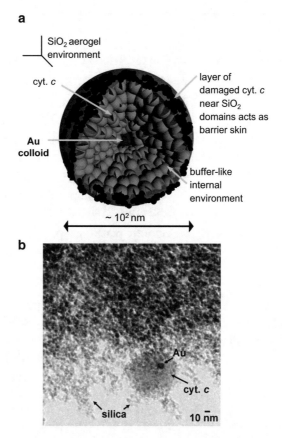

a

SiO$_2$ aerogel environment

cyt. *c*

layer of damaged cyt. *c* near SiO$_2$ domains acts as barrier skin

Au colloid

buffer-like internal environment

$\sim 10^2$ nm

b

Au

cyt. *c*

silica

10 nm

Fig. 1 (**a**) Proposed model of gold nanoparticle–nucleated protein–protein super-structure showing a specifically adsorbed protein monomolecular layer that initiates self-organization of free protein into a disordered superstructure. The low radius of curvature of the initial Au–protein structure facilitates molecular association and packing. Geometric estimates for packing shells of spheres about a nucleating sphere indicate that at a radii ratio, $r_{Au}/r_{cyt.c}$, of 20–30 (>70-nm Au), the Au surface should look planar to the protein. (**b**) Transmission electron micrograph of Au~cyt.*c* superstructure within Au~cyt.*c*–SiO$_2$ composite aerogel in which the silica domains of the aerogel, the protein superstructure, and the nucleating gold colloid are distinguishable. (Reproduced from ref. [17] with permission from the American Chemical Society, copyright 2003)

nanoparticles [24, 25]. The encapsulated cyt. *c* retains structural stability and gas-phase activity for nitric oxide, though self-organization into Au~cyt.*c* superstructures makes the protein slightly more stable within the aerogels. Some charge-induced agglomeration may occur even when cyt.*c* is encapsulated alone thereby reducing contact with the silica framework for a large fraction of the protein. We describe herein how to prepare both types of bioaerogels, those containing Au~cyt.*c* superstructures and those containing cyt.*c* without metal nanoparticles.

2 Materials

2.1 Silica Sol Preparation

1. Polypropylene disposable beakers (50 mL, Fisher).
2. Methanol (*see* **Note 1**).
3. Tetramethoxysilane (TMOS; also referred to as tetramethyl orthosilicate (98%, Aldrich)).
4. Ammonium hydroxide solution (ACS reagent, 28.0–30.0% NH_3 basis, Sigma-Aldrich).
5. General purpose polypropylene scintillation vials (16 mm×57 mm, volume size 6.5 mL, Sigma-Aldrich) with bottom end sliced off.
6. Generic plastic wrap.
7. Parafilm M™ laboratory wrapping film.

2.2 Au~cyt.c Superstructure Suspension or cyt.c Solution Preparation

1. Phosphate buffer: 50 mM, pH 7, made from sodium phosphate, monobasic or potassium phosphate, monobasic and sodium phosphate, dibasic, anhydrous or potassium phosphate, dibasic, anhydrous.
2. Cytochrome *c* (cyt. *c*) solution: 0.7–0.9 mM in prepared phosphate buffer, made from cyt.*c* from equine heart (≥95% based on Mol. Wt. 12,384, Sigma-Aldrich) used as received and stored at –20 °C. Once prepared, all buffered solutions are stored at 4 °C (*see* **Note 2**).
3. Colloidal gold sol, with a suspension density of 1.3×10^{13} Au particles mL^{-1}, made from commercial colloidal 5-nm gold sol (5×10^{13} Au particles mL^{-1}) or commercial colloidal 10-nm gold sol (5.7×10^{12} Au particles mL^{-1}) (20 mL, stored in the dark at 4 °C, BBI Solutions) (*see* **Note 3**).
4. Disposable cuvette or quartz cuvette (1-cm path length).
5. Glass 20-mL scintillation vials (O.D.×Height (with cap): 28 mm×61 mm, Wheaton).

2.3 Au~cyt.c or cyt.c Silica Sol–Gel Preparation

1. Polypropylene disposable beakers (50 mL, Fisher).
2. Ethanol, absolute, 200 proof, 99.5% A.C.S. reagent.
3. Acetone, HPLC grade.
4. Plastic syringe plunger.
5. Glass 20-mL Scintillation Vials (O.D.×Height (with cap): 28 mm×61 mm, Wheaton).

2.4 Au~cyt.c or cyt.c Silica Aerogel Preparation and UV–Visible Spectroscopy of Au~cyt.c–Silica or cyt.c–Silica Aerogels

1. Carbon dioxide, siphon-tube cylinder.
2. Au~cyt.*c*–silica or cyt.*c*–silica aerogel.
3. Cardboard platform (*see* **Note 4**).

2.5 Transmission Electron Microscopy of Au~cyt.c–Silica Aerogels

1. Au~cyt.*c*–silica aerogel.

2. PELCO® Center-Marked Grids, 200 mesh, 3.0-mm O.D., Copper (Ted Pella, Inc.).

3. Coat-Quick "G" grid-coating pen (Ted Pella, Inc.).

2.6 Nitric Oxide Sensing of Au~cyt.c– or cyt.c–Silica Aerogels

1. Au~cyt.*c*– or cyt.*c*–silica aerogel.

2. Argon (or other inert gas), cylinder.

3. Small cylinder (i.e., 4 L) of 10% NO in argon (prepared in-house) (*see* **Note 5**).

4. Tygon tubing (or other NO-resistant tubing), "T" switch valve, and syringe needles.

5. Disposable cuvette (Sarstedt, Acryl, 10 mm × 10 mm × 55 mm) with rubber septum cap.

3 Methods

One method that ensures that the heme protein, cyt.*c* remains viable throughout the conditions necessary to form an aerogel is to first associate it with nucleating gold or silver nanoparticles. Nucleation of a Au~cyt.*c* superstructure is initiated when the positively charged lysines at the heme edge of the protein specifically adsorb to the negatively charged, citrate-stabilized colloidal gold surface [17, 26–28]. This ordered, highly curved, nanometric surface presents the negatively charged side of the protein to the medium in such a way that rapid (and likely weak) protein–protein association organizes the non-adsorbed protein into the multilayered superstructure (Fig. 1a) [17, 29]. The organization of the superstructures is rapid and occurs in <0.1 s upon addition with stirring of an aliquot of buffered cyt.*c* to an aliquot of colloidal gold (or silver) sol [17, 18]. These superstructures are captured in the silica wet gel as isolated species, survive within the SCF-processed aerogel, and have been verified by transmission electron microscopy. The TEM-observed size of an aerogel-incorporated Au~cyt.*c* superstructure is consistent with that estimated from the projected size of metal nanoparticle–nucleated multilayers of ~3.4-nm globular cyt.*c* containing the stoichiometry of colloidal particles to protein molecules present in the original medium (Fig. 1b).

Alternately cyt.*c* can remain viable when encapsulated in sol-derived gels that are processed to form aerogels in the absence of metal nanoparticles when the cyt.*c* concentration within the aerogel remains within 5–15 μM and when the phosphate buffer concentration of the originating cyt. *c* solution is around 40 mM [24, 25]. Higher concentrations of cyt.*c* in buffer do not result in proportionately increased cyt.*c* visible absorbance signals within the protein–silica composite gel but do result in greater scattering of the optical signal. Three specific buffer concentrations, 4.4, 40, and 70 mM were systematically studied to make solutions of cyt.*c* for encapsulation. The mid-buffer strength resulted in aerogel-incorporated cyt.*c* remaining most stabilized to unfolding as monitored using the visible absorbance of the protein [24]. A high buffer strength is important to provide resistance to pH changes. However, the buffer strength should not be so high that cyt.*c* proteins are completely electrostatically shielded from each other, preventing any potential stabilizing protein–protein association within the aerogels. When cyt.*c* is gel-encapsulated in the absence of Au or Ag nanoparticles in aerogels, the absence of a high Z nanoparticulate core makes TEM analysis less effective.

Confirming cyt.*c* activity within aerogels is accomplished by visible spectroscopy, whether the protein is encapsulated alone or as Au~cyt.*c* superstructures. The translucent nature of the protein–silica composite aerogels makes it possible to measure absorbance of the characteristic Soret band of cyt.*c* within aerogels for comparison to the Soret absorbance and peak center of cyt.*c* in buffered solution. The protein within the aerogel also retains characteristic sensing activity as it can bind gas-phase nitric oxide as indicated by reversible shifts in the Soret peak position while toggling gas flow between argon and NO.

3.1 Silica Sol Preparation

1. Label two disposable 50-mL polypropylene beakers Beaker A and Beaker B. In Beaker A, add 1.88 g of tetramethoxysilane and 2.88 g of methanol, then cover the beaker with Parafilm. Place a magnetic stir bar in Beaker B, and add 0.75 g of water and 3.00 g of methanol. Cover the beaker in Parafilm.

2. Place Beaker B on a stir plate inside a fume hood and add 5 μL of ammonium hydroxide solution via syringe through the Parafilm while stirring. Immediately following the addition of ammonium hydroxide, add the contents of Beaker A to Beaker B. Stir the mixture for 20 min with Parafilm cover.

3. While the mixture is stirring, prepare the gel molds. Each mold consists of a polypropylene scintillation vial (with bottom cut off) and cap. Cover the cap end of the vial in plastic wrap, secure the cap on, and place the vials cap-end down on the bench top with opened ends facing up (*see* **Note 6**).

3.2 Au~cyt.c Superstructure Suspension or cyt.c Solution Preparation

1. A pH 7, sodium or potassium phosphate buffer is typically prepared in an ~750-mL portion for use over a period of 2 weeks. In the case of a 50 mM buffer, prepare 500 mL of 0.05 M monobasic salt and 500 mL of 0.05 M dibasic salt. Add aliquots of the monobasic salt solution to the dibasic salt solution until the pH is 7.00 as monitored using a pH electrode and meter.

2. Weigh out 0.023 g of cyt.*c* and dissolve in 2 mL of prepared sodium or potassium phosphate buffer.

3. Take 20 μL of cyt.*c* solution and dilute with 3 mL of buffer. Record UV–Vis spectra from 300 to 700 nm. Subtract the background absorbance at ~409 nm from the peak absorbance at ~409 nm to determine the protein absorbance at 409 nm. Use the extinction coefficient of 106,100 M^{-1} cm^{-1} to determine the concentration of the solution. Back calculate the concentration in the original solution. Typical concentrations of the original cyt.*c* solution range from 0.7 to 0.9 mM.

4. The concentration of cyt.*c* in the bioaerogel for maximum viability should be 5–15 μM. As long as the cyt.*c* concentration stays constant within the aerogel, the number of colloidal gold nanoparticles is less important when making Au~cyt.*c* superstructures for encapsulation. For a metal nanoparticle-to-protein ratio of 1:5700 and a final cyt.*c* concentration of 15 μM, the gold will need to be diluted. First, dilute the stock solution of 5-nm gold sol (with a suspension density of 5×10^{13} Au particles mL^{-1}) to 1.3×10^{13} Au particles mL^{-1} by adding 260 μL of the colloidal suspension to 740 μL of water with mixing.

5. About 10 min into stirring the silica sol (Subheading 3.1, **step 2**), synthesize the Au~cyt.*c* superstructures by slowly adding 117 μL of buffered 0.72 mM cyt.*c* solution dropwise to 683 μL of 1.3×10^{13} Au particles mL^{-1} colloidal gold suspension in a clean 20-mL glass scintillation vial while stirring on a stir plate with a stir bar (*see* **Note 7**). Stir the resulting nanoparticle–protein suspension for 10 min.

6. Alternatively, to make aerogels with cyt.*c* encapsulated alone, prepare an appropriately concentrated cyt.*c* solution by adding 117 μL of 0.72 mM prepared cyt.*c* solution (Subheading 3.2, **step 2**) together with 683 μL of prepared buffer (Subheading 3.2, **step 1**). Swirl to mix (*see* **Note 7**).

3.3 Au~cyt.c– or cyt.c–Silica Gel Preparation

1. Once the sol mixture has finished stirring for 20 min (Subheading 3.1, **step 2**), pipet 3 mL of this sol mixture into a clean polypropylene 50-mL beaker.

2. Add 500 μL of either the Au~cyt.*c* suspension (formed in Subheading 3.2, **step 5**) or the cyt.*c* solution (formed in Subheading 3.2, **step 6**) dropwise to the 3-mL sol mixture via a Pasteur pipette over the course of ~1 min. While adding either the Au~cyt.*c* suspension or the cyt.*c* solution, it is important to lightly stir or lightly shake the mixture to uniformly distribute protein throughout the sol, avoiding large red clumps from forming.

3. Pour approximately 0.5 mL of the resulting protein-doped sol into each mold to fill a total of six-to-seven molds. In addition, pour some of the "plain" silica sol into one or two molds to act

as a control during the supercritical drying process. Cover the sliced-open ends of the molds in Parafilm and leave to gel at 4 °C overnight.

4. After approximately 12 h of refrigeration, the protein-doped silica sols in the molds have gelled.

5. Remove the Parafilm, cap, and plastic wrap. Gently push the gels out of the molds using the circular disk end of a syringe plunger into clean 20-mL glass scintillation vials filled with ethanol. Store like gels (i.e., gels with the same ratio of gold to cyt.c or same concentration of cyt.c) together within one vial that is filled to the top with ethanol, capped, and stored at 4 °C.

6. Every 4 h during the first day, decant the ethanol off the gels and replace with fresh ethanol.

7. For the next 3 days, soak the wet sol gels in acetone, decanting and adding fresh acetone three times a day.

3.4 Au~cyt.c–Silica or cyt.c–Silica Aerogel Preparation

1. These instructions assume the use of a Polaron E3000 Series Critical Point Drying Apparatus. Attach this drying apparatus via Tygon tubing to an Isotemp Refrigerating Circulator (Fisher) and cool the apparatus initially so that the apparatus thermometer reads 10 °C, typically by setting the circulator temperature to 8 °C. Before the gels are inserted into the apparatus, a leak test must be performed to ensure that no carbon dioxide leaks from the chamber. To accomplish this step, fill the transfer boat with acetone, place it in the apparatus, seal the door, fill the apparatus with liquid carbon dioxide about half way, and listen for hissing at the doors and valves where O-rings or seals may be compromised. The affected O-rings or seals must be replaced if a leak is found.

2. Once the apparatus has been checked for leaks, open the drain valve, releasing the acetone and carbon dioxide through a Tygon tubing into a waste container held in a fume hood. Open the door of the apparatus and remove the transfer boat.

3. Carefully place the wet gels into the transfer boat's three long sections with forceps (no specimen baskets or gauze covers are used) and submerge them in acetone before placing the boat into the drying apparatus and closing the apparatus door. Typically, no more than eighteen 0.5-cm thick, 1-cm diameter gels can be fitted into the boat at one time.

4. Fill the apparatus with carbon dioxide and while maintaining the liquid CO_2 level above the samples, open the drain to release acetone for 5 min as soon as acetone can be seen draining to the bottom through the apparatus window (*see* **Note 8**). Close the drain and 5 min later, open the drain for 5 min again while maintaining the liquid CO_2 level above the samples. Repeat this process one more time. Then, open the drain under flowing $CO_2(l)$ for 5 min at a time every 20–45 min throughout the

course of at least 6 h for a total drainage time of approximately 45–50 min (*see* **Note 9**). It is important to monitor the level of fluid in the container by visible observation through the front window of the dryer to ensure that the level does not drop below the top of the boat during draining—the gels must remain submerged throughout the rinsing process.

5. When replacement of acetone within the wet gels by liquid carbon dioxide is complete, drain the carbon dioxide level to just above the prongs on the boat. Then raise the circulator temperature to 40 °C to bring liquid carbon dioxide above its critical temperature and pressure ($T_c = 31$ °C; $P_c = 7.4$ MPa). After equilibration at the critical temperature and pressure, typically 15–20 min, open the vent valve to release the supercritical fluid over the course of approximately 45 min (*see* **Note 10**).

6. When the pressure gauge drops to zero, open the apparatus door and remove newly dried aerogels and place in clean sample vials with forceps, being careful not to squeeze the gels too hard as they are brittle and will crack easily. The silica gels that are dried along with the Au~cyt.*c* or cyt.*c* gels are used as a visual reference to check that full carbon dioxide solvent transfer occurs. If these silica gels as well as the Au~cyt.*c* or cyt.*c* gels look cloudy in the middle, then full solvent replacement did not occur; if any acetone remains in the middle of the gels, pore collapse will result during drying.

3.5 UV–Visible Spectroscopy of Au~cyt.c–Silica or cyt.c–Silica Aerogels

1. These instructions assume the use of a single-beam spectrophotometer such as an Agilent model 8453 photodiode single-array spectrophotometer. Secure a cardboard platform to support the 5-mm thick, 1-cm diameter aerogel (*see* **Note 4**) inside the spectrometer sample compartment in the path of the beam. Scan an air background with this platform in place.

2. Before placing the aerogels onto the platform, measure the exact path length of the gels (thickness) with a micrometer.

3. Record a UV–visible spectrum from 300 to 800 nm for one Au~cyt.*c*–silica or cyt.*c*–silica aerogel. Perform a background subtraction (due to the large rising background from the silica gel) to determine the peak height, peak center, and peak width of the Soret peak of the aerogel-incorporated heme protein. Use the extinction coefficient of $106,100$ M^{-1} cm^{-1} and the measured thickness of the gel (path length) to determine the concentration of cyt.*c* in the aerogel and compare this value to that which was added (15 μM) to determine the fraction of cyt.*c* that remains viable within the aerogel. Examples of the UV–Vis spectra for Au~cyt.*c*–silica and cyt.*c*–silica aerogels are shown in Fig. 2b, c, respectively.

3.6 Transmission Electron Microscopy of Au~cyt.c–Silica Aerogels

1. These instructions assume the use of a JEOL 2010F field-emission transmission electron microscope equipped with a Gatan 794 MSC CCS camera. Extract thin slices of the Au~cyt.c–silica aerogel from the interior of the aerogel monolith (*see* **Note 11**). This step avoids sample preparation artifacts associated with solvent and grinding and allows one to obtain a micrographic image that captures all three components (gold nanoparticle, cyt.c, and silica aerogel), although the cyt.c and silica are not easy to distinguish unless the superstructure is sited towards the edge of the slice as in Fig. 1b.

2. Mount the submillimeter flakes of Au~cyt.c–silica aerogel directly to a copper-200-mesh TEM grid with the aid of Coat-Quick "G," an adhesive grid-coating pen.

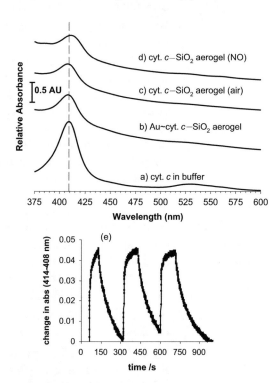

Fig. 2 Visible spectra of 15 μM cytochrome *c* in (**a**) 50 mM phosphate buffer solution; (**b**) $Au_{(5-nm)}$~cyt.c–SiO_2 aerogel; (**c**) cyt.c–SiO_2 aerogel (exposed to air); (**d**) cyt.c–SiO_2 aerogel (exposed to nitric oxide for 3.5 min). These representative spectra of each type of composite aerogel are offset for clarity; the *dashed line* denotes the position of the Soret peak of cyt.c in buffer. While each spectrum is derived for media containing 15 μM cyt.c, the thickness of the various aerogels is only 0.2–0.5 cm compared to cuvette's 1-cm path length for the protein in solution, which results in a higher unnormalized absorbance. Monitoring (at 414 nm) for the protein encapsulated within a Au~cyt.c–SiO_2 composite aerogel architecture demonstrates the reversible shift in the Soret intensity as the gas flow is toggled between argon and NO (**e**). (Reproduced from refs. [17] and [24] with permission from the American Chemical Society, Copyright 2003 and 2012)

3. Obtain the TEM images at low electron-beam fluence in order to minimize electron-beam damage to the sample.

3.7   Nitric Oxide Sensing of Au~cyt.c– Silica or cyt.c–Silica Aerogels

1. These instructions assume the use of an Agilent model 8453 photodiode single-array spectrophotometer, although this procedure could be adapted to other spectrophotometers. In a well-ventilated fume hood, place the spectrometer and the secured tank of 10 % NO/argon (*see* **Note 5**). Connect the Tygon tubing to both the argon and 10 % NO/argon gas tanks and attach the ends of the tubing to a T-valve. This arrangement allows a simple turn of the valve to alternate between the two gas flows.

2. Place the cuvette in the spectrometer and run as a blank sample, from 350 to 700 nm.

3. Measure the path length or aerogel thickness with a micrometer. Place the Au~cyt.*c*–silica or cyt.*c*–silica aerogel in the cuvette. Cap the cuvette with a rubber septum and insert two syringe needles: one for inflow of either the argon or the NO–argon mixture and the other as an exhaust. Obtain spectra to ensure the aerogel is properly placed in the cuvette such that it lies in the beam path.

4. Purge the aerogel and cuvette with a low flow rate of argon (a high flow rate is too turbulent and will move the aerogel out of the beam path). After an ~10-min purge, obtain the UV–visible spectrum, noting the position of the Soret band; Fig. 2b, c show examples, respectively, of Au~cyt.*c* and cyt.*c* spectra with the Soret at ~408 nm.

5. Turning the T-valve, switch to the 10 % NO–argon flow and obtain a spectrum at a series of timed intervals (~3–10 min), again noting the position of the Soret band and the splitting of the side bands (*see* Fig. 2d). Upon exposure to NO the Soret will shift to ~414 nm.

6. Switch back to argon flow and again obtain the UV–visible spectra. If the wavelength maximum of the Soret band returns to 408 nm, the binding of gas-phase NO is a reversible process.

7. With this experimental setup, the kinetics of the NO binding can be monitored. Using the Agilent software, measure the peaks at 408 and 414 nm every second over a given time span. Flow argon first to obtain a baseline, then switch the flow to NO–argon, then back to argon at timed intervals. Repeat this process, alternating between the two gases to test the time for the NO-binding response. Plot the change in absorbance between 414 and 408 nm as a function of time to yield information about the time it takes for the aerogel-encapsulated cyt.*c* to react with gas-phase NO; *see* Fig. 2e.

4 Notes

1. Due to the hygroscopic nature of methanol, tetramethoxysilane, and ammonium hydroxide solution, these reactants and reagents should be replaced every 1–2 months to avoid working with solutions that have become wet. The increase in water content over the as-prepared or as-received liquids will affect the sol-to-gel transition time and the final gel structure.

2. The water used in all experiments is 18 MΩ cm (Barnstead NANOpure™).

3. Colloidal gold is purchased in small quantities (20 mL) and only used if the suspension remains clear with no precipitate. Once precipitate forms, a new suspension must be purchased. Although commercial colloidal gold is used for size consistency in this procedure, it is also possible to synthesize citrate-stabilized gold colloids of varying sizes following the procedure initially developed by Turkevich and coworkers [30]. Silver colloids have also been synthesized and used in place of colloidal gold for this protocol [18].

4. A typical platform is prepared by cutting out a 2.5 cm × 2.5 cm piece of cereal box cardboard, folding it in half, cutting halfway up on the fold, then folding the two pieces created by cutting up. Open the folds and tape the flaps against the spectrometer sample compartment so that a small bent surface exists in which to position an aerogel directly in front of the beam path.

5. The experiment is performed in the fume hood to limit exposure to NO gas because sustained levels of NO may result in direct tissue toxicity. This precaution includes placing the spectrometer, cuvette, and gas cylinder in the hood as NO can also react with water to produce heat and corrosive fumes. NOTE: *on contact with air, NO will immediately convert to the highly poisonous nitrogen dioxide, nitrogen tetroxide, or both.*

6. The plastic wrap ensures that when sol is poured into the mold, a flat gel surface will form. When securing the cap onto the vial, the plastic wrap must not be pulled too tightly because the plastic wrap may sever and allow sol to flow through into the cap.

7. The exact volumes used will vary as the concentration of original cyt.*c* solution varies between 0.7 and 0.9 mM.

8. When carbon dioxide is added to the apparatus, it will initially be in the gas form, but condense to liquid immediately. The acetone will begin to seep down to the bottom of the apparatus as it mixes with the carbon dioxide and this mixing can be seen through the apparatus window. This liquid–liquid phase separation will occur before the apparatus is completely full of

carbon dioxide, so the drain should be opened as soon as acetone is seen seeping to the bottom. However, the rate of filling with carbon dioxide needs to be higher than the rate of draining such that the apparatus remains completely full even while the drain is open.

9. The carbon dioxide/acetone mixture that drains out of the apparatus will flow out of the apparatus so quickly that the drain tube will freeze stiff with ice collecting on the outside. In the first few rinses when acetone is the predominant component, the acetone (and water, since the acetone is not anhydrous) may cause solid frozen pieces to clog the drain tube thereby building up pressure in the drain tube, risking a violent popping of the tube off of the apparatus in the worst case. Such a pressure buildup should be avoided by watching for clogs and listening for a stoppage of flow. If a clog is detected, the drain should be closed for a minute to allow the clog to melt. Throughout the course of the day, as more rinsing takes place, the effluent from the drain tube will increasingly resemble dry ice, with no wet ice forming, and will smell less of acetone. Typically 45–50 min of drainage over the course of 6 h ensures complete solvent transfer of carbon dioxide into the interior of the gels, but strong olfactory evidence that no acetone is present in the drain effluent can also be used as another signal that complete transfer into carbon dioxide has occurred.

10. Raising the temperature inside the dryer above the critical temperature and pressure of carbon dioxide typically takes ~30 min. The transition from liquid to supercritical fluid can be observed as the liquid line above the prongs of the boat disappears. After this meniscus has disappeared, equilibration typically takes 15–20 min. Releasing the supercritical fluid through the vent valve too quickly may cause a decrease in the cyt.*c* viability within the aerogel as gauged by a decrease in the optical density of the Soret absorbance as observed by UV–Vis spectroscopic analysis; such degradation has been noted when releasing the fluid in ~10 min.

11. A TEM analysis is not effective for cyt.*c*–silica aerogels because of the absence of high Z contrast provided by the Au or Ag nanoparticles.

Acknowledgements

The authors gratefully acknowledge the support of this work by the U.S. Office of Naval Research and our colleague, Michael Doescher, for the graphics development of the model in Fig. 1. A.S.H. was an NRC–NRL Postdoctoral Associate (2004–2006). J.M.W. was an NRC–NRL Postdoctoral Associate (2000–2004).

References

1. Hitihami-Mudiyanselage A, Senevirathne K, Brock SL (2013) Assembly of phosphide nanocrystals into porous networks: formation of InP gels and aerogels. ACS Nano 7:1163–1170

2. Fricke J (1986) Aerogels. Springer, Berlin

3. Hüsing N, Schubert U (1998) Aerogels—airy materials: chemistry, structure, and properties. Angew Chem Int Ed 37:22–45

4. Aegerter AM, Leventis N, Koebel MM (eds) (2011) Aerogels handbook. Springer, New York, NY

5. Leventis N, Elder IA, Anderson ML, Rolison DR, Merzbacher CI (1999) Durable modification of silica aerogel monoliths with fluorescent 2,7-diazapyrenium moieties. Sensing oxygen near the speed of open-air diffusion. Chem Mater 11:2837–2845

6. Plata DL, Briones YJ, Wolfe RL, Carroll MK, Bakrania SD, Mandel SG, Anderson AM (2004) Aerogel-platform optical sensors for oxygen gas. J Non Cryst Solids 350:326–335

7. Rolison DR, Pietron JJ, Long JW (2009) Controlling the sensitivity, specificity, and time signature of sensors through architectural design on the nanoscale. ECS Transactions 19:171–179

8. Carroll MK, Anderson AM (2011) Aerogels as platforms for chemical sensors. In: Aegerter AM, Leventis N, Koebel MM (eds) Aerogels handbook, Springer. New York, NY, pp 637–650

9. Rolison DR (2003) Catalytic nanoarchitectures—the importance of nothing and the unimportance of periodicity. Science 299:1698–1701

10. Pietron JJ, Stroud RM, Rolison DR (2002) Using three dimensions in catalytic mesoporous nanoarchitectures. Nano Lett 2:545–549

11. Anderson ML, Morris CA, Stroud RM, Merzbacher CI, Rolison DR (1999) Colloidal gold aerogels: preparation, properties, and characterization. Langmuir 15:674–681

12. Anderson ML, Stroud RM, Rolison DR (2002) Enhancing the activity of fuel-cell reactions by designing three-dimensional nanostructured architectures: catalyst-modified carbon–silica composite aerogels. Nano Lett 2:235–240, correction: (2003) Nano Letters 3, 1321

13. Chervin CN, Ko JS, Miller BW, Dudek L, Mansour AN, Donakowski MD, Brintlinger T, Gogotsi P, Chattopadhyay S, Shibata T, Parker JF, Hahn BP, Rolison DR, Long JW (2015) Defective by design: vanadium-substituted iron oxide nanoarchitectures as cation-insertion hosts for electrochemical charge storage. J Mater Chem A 3:12059–12068

14. Maury S, Buisson P, Pierre AC (2001) Porous texture modification of silica aerogels in liquid media and its effect on the activity of a lipase. Langmuir 17:6443–6446

15. Innerlohinger J, Weber HK, Kraft G (2006) Aerocellulose: aerogels and aerogel-like materials made from cellulose. Macromol Symp 244:126–135

16. Harper-Leatherman AS, Wallace JM, Long JW, Rhodes CP, Pettigrew KA, Graffam ME, Abunar BH, Iftikhar M, Ndoi A, Capecelatro AN, Rolison DR (2016) Redox cycling within nanoparticle-nucleated self-organized protein superstructures: Electron transfer between gold nanoparticle cores, tannic acid, and cytochrome *c*. Manuscript in preparation

17. Wallace JM, Rice JK, Pietron JJ, Stroud RM, Long JW, Rolison DR (2003) Silica nanoarchitectures incorporating self-organized protein superstructures with gas-phase bioactivity. Nano Lett 3:1463–1467

18. Wallace JM, Dening BM, Eden KB, Stroud RM, Long JW, Rolison DR (2004) Silver-colloid-nucleated cytochrome *c* superstructures encapsulated in silica nanoarchitectures. Langmuir 20:9276–9281

19. Wallace JM, Stroud RM, Pietron JJ, Long JW, Rolison DR (2004) The effect of particle size and protein content on nanoparticle-gold–nucleated cytochrome *c* superstructures encapsulated in silica nanoarchitectures. J Non Cryst Solids 350:31–38

20. Lynch I, Cedervall T, Lundqvist M, Cabaleiro-Lago C, Linse S, Dawson KA (2007) The nanoparticle-protein complex as a biological entity: a complex fluids and surface science challenge for the 21st century. Adv Colloid Interface Sci 134–135:167–174

21. Cedervall T, Lynch I, Lindman S, Berggard T, Thulin E, Nilsson H, Dawson KA, Linse S (2007) Understanding the nanoparticle–protein corona using methods to quantify exchange rates and affinities of proteins for nanoparticles. Proc Natl Acad Sci 104:2050–2055

22. Del Pino P, Pelaz B, Zhang Q, Maffree P, Ninehaus U, Parak WJ (2014) Protein corona formation around nanoparticles—from the past to the future. Mater Horizon 1:301–313

23. Mahmoudi M, Lynch I, Ejtehadi R, Monopoli MP, Bombelli FB, Laurent S (2011) Protein-nanoparticle interactions: opportunities and challenges. Chem Rev 111:5610–5637

24. Harper-Leatherman AS, Iftikhar M, Ndoi A, Scappaticci SJ, Lisi GP, Buzard KL, Garvey EM (2012) Simplified procedure for encapsulating cytochrome *c* in silica aerogel nanoarchitectures while retaining gas-phase bioactivity. Langmuir 28:14756–14765

25. Harper-Leatherman AS, Pacer ER, Kosciuszek ND (2016) Encapsulating cytochrome *c* in silica aerogel nanoarchitectures without metal nanoparticles while retaining gas-phase bioactivity. J Vis Exp 109:e53802

26. Bowden EF, Hawkridge FM, Blount HN (1984) Interfacial electrochemistry of cytochrome *c* at tin oxide, indium oxide, gold, and platinum electrodes. J Electroanal Chem Interf Electrochem 161:355–376

27. Kang CH, Brautigan DL, Osheroff N, Margoliash E (1978) Definition of cytochrome *c* binding domains by chemical modification. Reaction of carboxydinitrophenyl- and trinitrophenyl-cytochromes c with baker's yeast cytochrome c peroxidase. J Biol Chem 253:6502–6510

28. Keating CD, Kovaleski KM, Natan MJ (1998) Protein : colloid conjugates for surface enhanced Raman scattering: stability and control of protein orientation. J Phys Chem B 102:9404–9413

29. Koppenol WH, Margoliash E (1982) The asymmetric distribution of charges on the surface of horse cytochrome *c*. Functional implications. J Biol Chem 257:4426–4437

30. Turkevich J, Stevenson PC, Hillier J (1951) A study of the nucleation and growth processes in the synthesis of colloidal gold. Discuss Faraday Soc 11:55–75

Chapter 13

Enzyme Immobilization and Mediation with Osmium Redox Polymers

Gaige R. VandeZande, Jasmine M. Olvany, Julia L. Rutherford, and Michelle Rasmussen

Abstract

Enzymatic electrodes are becoming increasingly common for energy production and sensing applications. Research over the past several decades has addressed a major issue that can occur when using these biocatalysts, i.e., slow heterogeneous electron transfer, by incorporation of a redox active species to act as an electron shuttle. There are several advantages to immobilizing both the enzyme and mediator at the enzyme surface, including increased electron transfer rates, decreased enzyme leaching, and minimized diffusion limitations. Redox polymers consisting of a redox active center attached to a polymer backbone are a particularly attractive option because they have high self-exchange rates for electron transfer and tunable redox potential. Osmium (Os) polymers are the most well studied of this type of polymer for bioelectrocatalysis. Here, we describe the methods to synthesize one of the most common Os redox polymers and how it can be used to fabricate glucose oxidase electrodes. Procedures are also outlined for evaluating the enzymatic electrodes.

Key words Enzyme immobilization, Redox polymer, Osmium polymer, Mediated electron transfer, Mediator

---

1 Introduction

The use of enzymatic electrodes has increased dramatically over the past few decades due to the high specificity and relatively low cost of enzymes compared to many other types of catalysts, specifically metals such as platinum. Enzymes are frequently used for energy production devices like biofuel cells as well as for environmental or medical sensing. Initial bioelectrocatalysis research was performed in solution, but since then it has been shown that higher currents and greater stability are possible when the enzyme is appropriately immobilized.

1.1 Mechanisms of Electron Transfer

One important aspect of enzymatic electrodes is the mechanism of electron transfer: mediated vs. direct. Direct electron transfer (DET), where electrons are passed directly either to or from the active site of the enzyme, typically through a redox cofactor, is the

Shelley D. Minteer (ed.), *Enzyme Stabilization and Immobilization: Methods and Protocols*, Methods in Molecular Biology, vol. 1504, DOI 10.1007/978-1-4939-6499-4_13, © Springer Science+Business Media New York 2017

simplest option, because it does not require any other components. Figure 1a shows the electron transfer scheme for a DET anode. The enzymatic oxidation reaction occurs and electrons are transferred to the cofactor in **step 1**. The reduced cofactor is then oxidized back to its original state by the electrode in **step 2** and the enzyme is able to continue catalyzing the reaction. However, many enzymes are unable to perform direct electron transfer, because their cofactor or active site is too buried within the protein structure. Additionally, enzymes that can perform DET are limited to the enzymes that are close to the surface and require high surface area to produce large currents.

The alternative option is incorporation of an artificial intermediate redox species which can shuttle electrons to or from the enzyme's active site for mediated electron transfer (MET). The mediator takes the place of one of the enzyme's natural substrates and acts as either an electron donor or acceptor depending on whether the reaction is an oxidation or reduction. Figure 1b shows the electron transfer scheme for an anode using MET which occurs in four steps. The first step is the same as DET, when the enzymatic oxidation reaction occurs, and electrons are transferred to the cofactor. In **step 2**, the reduced cofactor then transfers electrons to a nearby mediator. The reduced mediator is able to pass the electrons to another neighbor in **step 3** which can repeat until one of the reduced mediators is close enough to the electrode to be oxidized

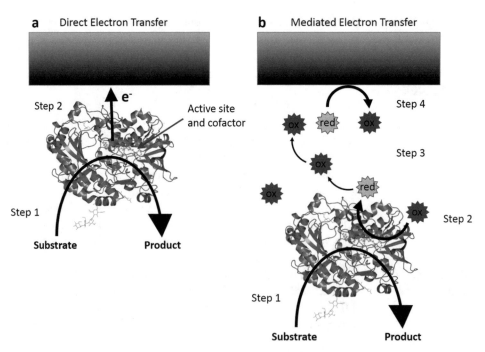

Fig. 1 Electron transfer at an enzymatic electrode using either: (**a**) direct electron transfer or (**b**) mediated electron transfer

in **step 4**. MET makes it possible to produce significantly larger current densities and can prevent electron loss from reduction of O_2, but it also makes the electrode more complicated, increasing the components needed and possibly affecting the stability.

1.2   Types of Mediator

There are several important aspects to consider when choosing a mediator [1]:

1. Reversibility: the chosen mediator must be reversible to avoid inactivation of the electrode.

2. Stability: the mediator should be stable in both its oxidized and reduced forms to prevent inactivation of the electrode.

3. Appropriate redox potential: the mediator redox potential should be similar to that of the redox potential of the enzyme (or its cofactor) to minimize the added overpotential and also to prevent undesirable reactions.

4. Rapid electron transfer: the mediator should be capable of fast electron transfer to or from the enzyme cofactor and also to or from the electrode.

In the earliest work with mediated electron transfer, the redox species was added directly to the solution during electrochemical experiments. The compound could freely diffuse between the enzyme and electrode. For example, Yahiro et al. showed that both glucose oxidase (GOx) and D-amino acid oxidase produced a potential when iron was used as a mediator [2]. Many other compounds have been used for mediation in solution, including methylene blue, benzoquinone, 2,2′-azino-bis(3-ethylbenzothiazoline-6-sulfonic acid) (ABTS), and methyl viologen [3–7]. There are several disadvantages to using the redox species in solution. The molecules may need to diffuse quite far to transfer electrons from the enzyme to the electrode, limiting the current produced. The cost tends to be higher as well because more mediator is needed each time the solution is replaced. In biofuel cell applications, a free mediator also makes it necessary to use a separator to prevent short-circuiting of the system. Immobilizing the redox species on the electrode addresses all of these problems.

A straightforward method to immobilize a mediator is to covalently attach it to the electrode surface on a self-assembled monolayer (SAM). Enzymatic electrodes have been reported with bound cytochrome c [8] or pyrroloquinoline quinone [9] covalently tethered to the surface. These electrodes showed bioelectrocatalytic activity with GOx and cytochrome c oxidase, respectively. The major disadvantage to this method is that immobilizing on a SAM limits the amount of mediator to that which can fit in a single monolayer. Increasing the surface area by increasing the electrode size is the only simple method of improving the performance of this type of electrode.

Electropolymerization of species to produce conductive polymer films is another method of immobilized mediation which has the additional advantage of also binding the enzyme, either by entrapment or covalent attachment. Early work in this area focused on entrapping GOx in a variety of polypyrrole derivatives [10–12]. The field was later expanded to include polymers such as polyaniline [13] and polythiophene [14] for use with a variety of enzymes.

The majority of work with immobilized redox mediators has been focused on organometallic hydrogel polymers. In these materials, a redox-active metal center is covalently attached to a polymer backbone such as poly(vinylpyridine) (PVP), poly(vinylimidazole) (PVI), or poly(ethyleneimine) (PEI). For immobilization of biomolecules, the polymer is cross-linked and the resulting polymer network swells when placed in solution, leading to rapid substrate diffusion and "wiring" of the enzyme to the electrode through the redox centers. The most commonly studied redox polymers contain Os centers, but work has also been done with polymers containing ruthenium [15] or ferrocene species [16–18].

1.3 Osmium Redox Polymers

The earliest work with Os redox polymers for enzymatic electrodes was done by Adam Heller in the 1980s. He reported a redox polymer containing bipyridinyl osmium complexes tethered to the backbone which mediated electron transfer from GOx (*see* Fig. 2) [19]. Further work was focused on adjusting the redox potential of the polymer by changing the ligands on the Os centers. A series of Os polymers has been synthesized which covers a wide range of potentials, making it possible to use the polymers with a large variety of enzymes at both cathodes and anodes [20–22] for applications such as biosensing [23–28] and implantable devices [29, 30].

The procedure outlined here includes the synthesis of one of the most common osmium redox complexes, Os(bpy)$_2$, and the subsequent attachment of the complex to a polymer backbone, PVI. Its use in electrode modification with GOx will be detailed and several methods for evaluating the modified electrode will be discussed, including electrochemical methods and determination of immobilized enzyme activity. While the procedures in this chapter are for a specific Os redox polymer and enzyme, these methods may be easily adapted for use with other polymers and enzymes.

2 Materials

2.1 Synthesis of PVI-Os(bpy)$_2$

2.1.1 Poly(1-Vinylimidazole) (PVI)

1. 1-Vinylimidazole.
2. 2,2′-azobis(isobutyronitrile) (AIBN).
3. Methanol.
4. Acetone.
5. Small round bottom flask (25 or 50 mL).
6. Stir plate and appropriately sized stir bar.

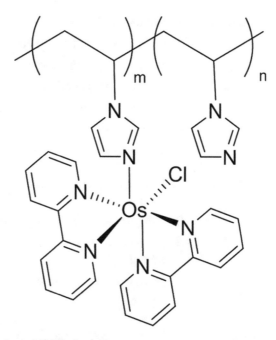

Fig. 2 Structure of PVI-Os(bpy)₂Cl

7. Filter and appropriately sized filter paper or filter with frit.

2.1.2 Bis(2,2'-Bipyridine-N,N')
Dichloroosmium(III)
Chloride Dihydrate,
[Os(bpy)₂Cl₂]Cl

1. Potassium hexachloroosmate (IV).
2. 2,2'-bipyridine.
3. Dimethylformamide (DMF).
4. Ethanol.
5. Reflux condenser.
6. Tubing to connect to condenser.
7. Stir plate and appropriately sized stir bar.
8. Small round bottom flask (25 mL).
9. Diethyl ether.
10. Filter and appropriately sized filter paper or filter with frit.

2.1.3 Bis(2,2'-Bipyridine-N,N')
Dichloroosmium(II),
[Os(bpy)₂Cl₂]

1. DMF.
2. Methanol.
3. Round bottom flask (50 mL).
4. Stir plate and appropriately sized stir bar.
5. Sodium hydrosulfite.
6. Ice.
7. Glass rod or spatula.
8. Diethyl ether.
9. Filter and appropriately sized filter paper or filter with frit.

2.1.4 PVI-Os(bpy)₂Cl	1. Ethanol.
	2. Round bottom flask (100 mL).
	3. Stir plate and appropriately sized stir bar.
	4. Reflux condenser.
	5. Tubing to connect to condenser.
	6. Diethyl ether.
	7. Rotary evaporator.

2.1.4   PVI-Os(bpy)$_2$Cl

1. Ethanol.
2. Round bottom flask (100 mL).
3. Stir plate and appropriately sized stir bar.
4. Reflux condenser.
5. Tubing to connect to condenser.
6. Diethyl ether.
7. Rotary evaporator.

2.2 Modification of Electrodes with Enzyme and Osmium Polymer

1. PVI-Os(bpy)$_2$Cl, produced as described (*see* Subheading 2.1): 10 mg/mL in ethanol.
2. Poly(ethylene glycol) diglycidyl ether (PEGDGE): 2.5 mg/mL.
3. Glucose oxidase (GOx) from *Aspergillus niger* (E.C. 1.1.3.4): 13 mg/mL.
4. Bovine serum albumin (BSA).
5. Glassy carbon (GC) electrodes, 3 mm diameter.

2.3 Determination of Immobilized Enzyme Activity

1. GOx-PVI-Os(bpy)$_2$Cl modified electrode, prepared as described (*see* Subheading 3.2).
2. 0.1 M phosphate buffer, pH 7.4.
3. D-glucose solution, prepared and allowed to mutarotate for 24 h (*see* **Note 1**).
4. 0.7 mM 2,6-dichlorophenolindophenol (DCPIP), prepared in water.
5. Small vial or beaker.
6. Visible spectrophotometer.
7. Cuvettes and caps.

2.4 Electrochemical Testing of Modified Electrodes

1. Potentiostat able to perform cyclic voltammetry and amperometry.
2. Stir plate and appropriately sized stir bar.
3. Small beaker or electrochemical cell.
4. Reference electrode suitable for aqueous experiments (typically Ag/AgCl or SCE).
5. Counter electrode (typically platinum mesh or wire, *see* **Note 2**).
6. GOx-PVI-Os(bpy)$_2$Cl modified electrode, prepared as described (*see* Subheading 3.2).
7. BSA-PVI-Os(bpy)$_2$Cl modified electrode, prepared as described (*see* Subheading 3.2).
8. 0.1 M phosphate buffer, pH 7.4.
9. 1 M D-glucose solution, prepared in phosphate buffer and allowed to mutarotate for 24 h (*see* **Note 1**).

3   Methods

3.1   Synthesis of PVI-Os(bpy)₂

The synthesis procedure described here is for one of the simplest osmium redox polymers [31–33]. It consists of a poly(1-vinylimidazole) backbone with tethered $Os(bpy)_2Cl_2$ groups as shown in Fig. 2. Different polymer backbones could be used, such as polyvinylpyridine, or alternative osmium centers could also be used, but it is possible that the procedures would need to be adjusted.

3.1.1   Poly(1-Vinylimidazole) (PVI)

1. Add 6 mL of 1-vinylimidazole to a small round bottom flask.
2. Heat to 70 °C under N_2.
3. While stirring, add 0.5 g AIBN.
4. Allow the reaction to proceed for 2 h. A yellow precipitate should form.
5. After cooling to room temperature, dissolve the precipitate in methanol.
6. Slowly add the PVI–methanol solution to rapidly stirred acetone.
7. Filter the pale yellow precipitate.

3.1.2   Bis(2,2′-Bipyridine-N,N′) Dichloroosmium(III) Chloride Dihydrate, [Os(bpy)₂Cl₂]Cl

1. To 15 mL of DMF in a small round bottom flask, add 0.16 g potassium hexachloroosmate (IV) and 0.125 g 2,2′-bipyridine.
2. Reflux for 1 h.
3. Cool to room temperature.
4. Filter the solution to remove the KCl that forms.
5. Add ~10 mL of ethanol.
6. The complex can be precipitated by adding ~200 mL diethyl ether while stirring.
7. Filter and wash the dark reddish-brown precipitate with diethyl ether.

3.1.3   Bis(2,2′-bipyridine-N,N′)dichloroosmium(II), [Os(bpy)₂Cl₂]

1. Dissolve 0.1 g $[Os(bpy)_2Cl_2]Cl$ in 30 mL of 2:1 mixture of DMF and methanol.
2. While stirring, slowly add a solution of sodium hydrosulfite (0.2 g in 20 mL of H_2O) dropwise over 30 min.
3. Cool the mixture on ice.
4. Scratch the walls with a glass rod or spatula to initiate crystallization.
5. Filter and wash with water, methanol, and diethyl ether.

3.1.4   PVI-Os(bpy)₂Cl

1. Dissolve 40 mg of $[Os(bpy)_2Cl_2]$ in 30 mL ethanol.
2. Reflux the solution for 30 min.

3. Add 70 mg of PVI dissolved in 10 mL ethanol.

4. Reflux the mixture for 72 h.

5. The redox polymer can be precipitated in diethyl ether and dried on a vacuum.

3.2 Modification of Electrodes with Enzyme and Osmium Polymer

Immobilizing enzymes on an electrode with a redox polymer is done by mixing with a cross-linker and dropcasting onto the surface. The cross-linker used here is able to react with amines and imidazoles, allowing for covalent binding of the enzyme and polymer, creating a 3D network. This procedure uses glucose oxidase with PVI-Os(bpy)$_2$Cl, but it could also be applied to other FAD-dependent enzymes or other redox polymers, provided that they contain free amine or imidazole groups for cross-linking.

1. In a small Eppendorf tube or glass vial, mix 72 μL of 10 mg/mL PVI-Os(bpy)$_2$Cl, 18 μL of 2.5 mg/mL PEGDGE, and 10 μL 13 mg/mL GOx. Vortex briefly to mix.

2. Apply 10 μL to each GC electrode.

3. Let dry in ambient conditions overnight.

4. For control experiments, make electrodes following the same procedure but substitute BSA for the GOx.

3.3 Determination of Immobilized Enzyme Activity

It is often times desirable to determine the activity of the enzyme when it is immobilized. By comparing the immobilized specific activity to that of the enzyme in solution, it is possible to determine whether the immobilization method affects the enzyme's ability to perform the reaction. Immobilization can decrease activity by limiting diffusion of the substrate or preventing conformation change of the enzyme. However, in some cases immobilization might increase the activity by improving the stabilization and providing a desirable environment for the enzyme. In the procedure described [34], the specific activity of the immobilized GOx is determined using the electron acceptor DCPIP which is blue in its oxidized form and colorless when reduced. This procedure can be used to determine the activity of GOx with any immobilization method.

1. In a small vial or beaker, mix 8.5 mL of phosphate buffer, 0.5 mL of DCPIP, and 1 mL of glucose solution.

2. Add 2 mL of the mixture to a cuvette.

3. Place the cuvette in the spectrophotometer and measure the absorbance at 555 nm.

4. Suspend the enzymatic electrode in the cuvette so that the modified surface is submerged. Leave the electrode in the solution for 5 min.

5. Remove the electrode, cover the cuvette with a lid, and mix by inverting 3–4 times.

6. Place the cuvette in the spectrophotometer and measure the absorbance at 555 nm.

The activity of the immobilized GOx can be calculated from the following equation:

$$\text{Enzyme activity}\left(\frac{\text{Units}}{\text{mg}}\right) = \frac{\left(A_{\text{initial}} - A_{\text{final}}\right)\left(2\,\text{mL}\right)}{\left(8.465\,\text{mM}^{-1}\text{cm}^{-1}\right)\left(5\,\text{min}\right)\left(1\,\text{cm}\right)\left(0.013\,\text{mg}\right)}$$

A_{initial} = initial absorbance at 555 nm.

A_{final} = absorbance at 555 nm after reaction with enzyme electrode.

2 mL = total volume of assay.

8.465 mM^{-1}cm^{-1} = millimolar extinction coefficient of DCPIP.

5 min = total time of reaction.

1 cm = path length of cuvette.

0.013 mg = mass of immobilized GOx.

One unit of GOx will oxidize 1 μm of β-D-glucose to D-gluconolactone per minute at room temperature at pH 7.4.

3.4 Electrochemical Testing of Modified Electrodes to Confirm Bioelectrocatalysis

Two types of electrochemical experiments are commonly performed to verify bioelectrocatalytic activity of enzymatic electrodes: cyclic voltammetry (CV) and amperometry. CV is a quick, straightforward experiment which will show whether the electrode is functioning as expected. Amperometry is a more sensitive method to quantify the electrode's activity.

3.4.1 Cyclic Voltammetry

In this technique, the potential of the working electrode is varied and the current is measured. Reversible redox systems will show an oxidation peak when scanning in the positive direction and a reduction peak when scanning in the negative direction. First the electrode is tested in a blank solution, typically buffer, containing none of the enzyme's substrate. Then the test is repeated in the presence of substrate. If the enzyme reaction is an oxidation (which is the case for the procedure described here), when substrate is added, the oxidation peak should increase and diffusional tailing (the drop off after the peak) should get smaller. The reduction peak should also get smaller.

1. In a small beaker or electrochemical cell, place the enzymatic electrode, reference electrode, and counter electrode in phosphate buffer. The volume of buffer will vary based on the size of the container but should be enough to cover the electrodes (*see* **Note 3**).

2. Program the potentiostat software with the following parameters (the names may vary based on brand of instrument):

   (a) Low potential: –0.1 V vs. Ag/AgCl.

   (b) High potential: 0.6 V vs. Ag/AgCl.

   (c) Scan rate: 5 mV/s (*see* **Note 4**).

   (d) Sweep segments or number of cycles: four segments or two cycles (*see* **Note 5**).

3. After programming the software with the above parameters, run two cycles with the electrode in buffer.

4. Add enough of the 1 M glucose solution to make the final concentration 100 mM glucose.

5. Run two more cycles using the same parameters.

A plot of the second cycle for each solution should confirm bioelectrocatalysis. As seen in Fig. 3, when no glucose is present, a typical cyclic voltammogram for the Os redox polymer is observed, with a redox potential of ~325 mV vs. Ag/AgCl at pH 7.4. In the presence of 100 mM glucose, the oxidation current increases rather dramatically and reaches a limiting current density close to 1 mA/cm^2. The enzyme is oxidizing glucose and FAD is being reduced to FADH$_2$. The FAD is being regenerated by passing its electrons to the Os centers. The reduced Os is then oxidized by the electrode, generating a large current. The reaction is no longer diffusion limited because the enzyme is continuously producing electrons at the electrode. The reduction peak at ~300 mV vs. Ag/AgCl is no longer observed because all of the Os centers are being reduced by the enzyme rather than the electrode.

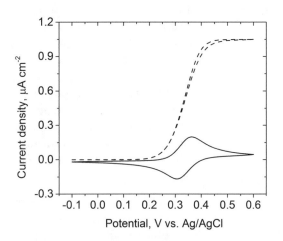

Fig. 3 Cyclic voltammograms of a PVI-Os(bpy)$_2$Cl and GOx modified electrode in 0.1 M pH 7.4 phosphate buffer containing 0 mM (*solid line*) and 100 mM (*dashed line*) glucose at a scan rate of 5 mV/s

For this experiment, a fixed potential is applied to the working electrode and the current is measured over time. The potential is chosen based on the CV. Typically the potential should be after the peak of interest (the oxidation peak for this procedure) by about 50–100 mV. The potential is applied and the current initially jumps to a large value before slowly decaying. It should eventually stabilize at a relatively constant value, i.e., steady state current. At this point, substrate can be injected into the solution. After each injection the current should be allowed to stabilize again. For this type of experiment, the solution needs to be stirred continuously but somewhat slowly to prevent noise in the data.

1. In a small beaker or electrochemical cell, place the enzymatic electrode, reference electrode, and counter electrode in phosphate buffer. The volume of buffer will vary based on the size of the container but should be enough to cover the electrodes. Make sure the stir bar is not too close to any of the electrodes (*see* **Note 6**).

2. Program the potentiostat software with the following parameters (the names may vary based on brand of instrument):

    (a) Potential: 0.5 V vs. Ag/AgCl.

    (b) Sample interval: how often the instrument takes a data point. Typically one point every second works well.

    (c) Run time: Sometimes these experiments take a long time so setting the run time to be long is better than too short. The experiment can always be stopped early but for most instruments the parameters cannot be changed after it is already running.

3. After programming the software with the appropriate parameters, begin the experiment.

4. Depending on the electrode, the amount of time it takes for the current to stabilize can vary greatly. Stabilization times of anywhere from a couple of minutes up to an hour are possible.

5. When the current has stabilized, begin injections of the 1 M glucose solution (*see* **Note 7**). After each injection allow the current to stabilize before continuing with the next. Again, this time may vary.

6. Continue with injections until the current shows no increase, indicating saturation of the enzyme. At this point, do 2–3 more injections and then stop the experiment.

7. Repeat the experiment with the control BSA-PVI-Os(bpy)$_2$Cl electrodes.

In a well-planned experiment, the amperometric data should look similar to that shown in Fig. 4a. After each injection of glucose, the current increases rapidly and then levels off, creating a

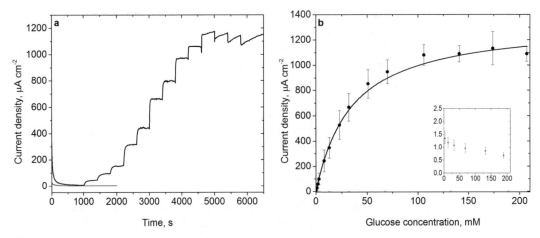

Fig. 4 (a) Amperometric data for a PVI-Os(bpy)$_2$Cl and GOx modified electrode in 0.1 M pH 7.4 phosphate buffer at 0.5 V vs. Ag/AgCl (black data points). Injections of a 1 M glucose solution were made with the following volumes sequentially: 10, 10, 10, 50, 50, 100, 100, 200, 200, 400, 400, 400, and 400 μL. The *red data points* show the results for control electrodes. **(b)** Plot of the steady state current density at each concentration from **(a)**. The parameters from fitting the data to the Michaelis–Menten equation are $i_{max} = 1340 \pm 50$ μA/cm^2 and $K_m = 35 \pm 4$ mM. *Inset:* Steady state current density measured at control electrodes containing BSA instead of GOx

staircase shaped curve. The results for the control experiments show no change in current with increasing concentration (*see* red line in Fig. 4a).

After determining the steady state current density at each concentration, a plot similar to that shown in Fig. 4b can be made, which shows a classic Michaelis–Menten type curve. At low concentrations, the current density increases linearly. At high concentrations, the current density change becomes smaller until eventually, at high enough concentrations, the enzyme is saturated and the current density remains constant regardless of increasing concentration. The inset in Fig. 4b shows the results for the control electrodes which has no response to increasing glucose concentration.

This data can be fit to the Michaelis–Menten equation to obtain V_{max} (or i_{max} in this case because the current density is a measure of the reaction rate) and K_m. These parameters are a good way of comparing electrodes made with different immobilization strategies using the same enzyme. The maximum current that can be produced, i_{max}, is a straightforward way of determining which electrode performs better. However, the Michaelis constant, K_m, is also useful, because it is an indicator of the enzyme's affinity for the substrate (glucose in this case). For the example shown, the i_{max} is 1340 ± 50 μA/cm^2 with a K_m of 35 ± 4 mM. This K_m is higher, but the same order of magnitude, as previously reported values for free enzyme at pH 7.4 [35], indicating that this immobilization method may slightly hinder the enzyme's ability to bind the substrate. The data for the control electrodes shows that without the GOx, there is no change in current in the presence of glucose.

4 Notes

1. Mutarotation is required because sugars have two stereoisomeric forms, α and β. In solution, glucose will interconvert between the two forms. Commercial D-glucose is typically the β form. Once added to solution, some of the glucose will convert to the α form until an equilibrium ratio is reached. Because glucose oxidase is only active for the β form, it is critical that the glucose solution is at equilibrium before it is used. If it is used prior to reaching equilibrium, the concentration of the substrate (β-D-glucose) will not be constant.

2. The counter electrode needs to have larger surface area than the working electrode, typically on the order of three times larger, and should be electrochemically inert. It is common to use platinum mesh wrapped at the end of platinum wire which makes a small electrode with large surface area.

3. It is critical that any conductive surfaces not come in contact with any other conductive surfaces. The reference electrode is sealed in glass so it should not be a problem. However, the counter electrode is entirely conductive so care should be taken to make sure it does not touch any metal on clamps, leads, etc.

4. The scan rate determines how fast the potentiostat moves from one potential to the next. For bioelectrochemistry, because reactions can be slow and also have multiple electron transfer steps, a slower scan rate is needed. Scan rates of 1–10 mV/s are typical.

5. The name of this parameter depends on the brand of instrument. Each cycle is two segments. At least two full cycles should always be performed because the first one is not representative. The surface of the electrode might go through some processes on the first scan that will not occur on any subsequent scans.

6. To obtain the best data in an amperometric experiment, the electrochemical cell setup and stirring need to be optimized. The working electrode should be suspended so that its surface is near the top of the solution. This minimizes the noise from stirring. The speed of stirring needs to be fast enough to rapidly mix the glucose when injected but not so fast that it creates noise in the current data.

7. Ideally the glucose should be injected near the stir bar to ensure rapid mixing. Inject the glucose slowly into the solution. If the solution is injected too quickly, the current may show a large increase initially before decreasing and eventually stabilizing at steady state.

References

1. Gallaway JW (2014) Mediated enzyme electrodes. Enzymatic Fuel Cells pp 146–180
2. Yahiro AT, Lee SM, Kimble DO (1964) Bioelectrochemistry. I. Enzyme utilizing biofuel cell studies. Biochim Biophys Acta 88(2):375–383
3. Janda P, Weber J (1991) Quinone-mediated glucose oxidase electrode with the enzyme immobilized in polypyrrole. J Electroanal Chem Interfacial Electrochem 300(1):119–127
4. Palmore G, Kim HH (1999) Electro-enzymic reduction of dioxygen to water in the cathode compartment of a biofuel cell. J Electroanal Chem 464(1):110–117
5. Tsujimura S, Tatsumi H, Ogawa J, Shimizu S, Kano K, Ikeda T (2001) Bioelectrocatalytic reduction of dioxygen to water at neutral pH using bilirubin oxidase as an enzyme and 2,2'-azinobis(3-3thylbenzothiazolin-6-sulfonate) as an electron transfer mediator. J Electroanal Chem 496:69–75
6. Zen J-M, Lo C-W (1996) A glucose sensor made of an enzymatic clay-modified electrode and methyl viologen mediator. Anal Chem 68(15):2635–2640
7. Qian J, Liu Y, Liu H, Yu T, Deng J (1996) An amperometric new methylene blue N-mediating sensor for hydrogen peroxide based on regenerated silk fibroin as an immobilization matrix for peroxidase. Anal Biochem 236(2):208–214
8. Katz E, Heleg-Shabtai V, Willner I, Rau HK, Haehnel W (1998) Surface reconstitution of a de novo synthesized hemoprotein for bioelectronic applications. Angew Chem Int Ed 37(23):3253–3256
9. Riklin A, Katz E, Wiliner I, Stocker A, Bückmann AF (1995) Improving enzyme electrode contacts by redox modification of cofactors. Nature 376(6542):672–675
10. Barlett P, Cooper J (1993) A review of the immobilization of enzymes in electropolymerized films. J Electroanal Chem 362(1):1–12
11. Schuhmann W (1995) Conducting polymer based amperometric enzyme electrodes. Microchim Acta 121(1):1–29
12. Cosnier S (1997) Electropolymerization of amphiphilic monomers for designing amperometric biosensors. Electroanalysis 9(12):894–902
13. Xu S, Minteer SD (2014) Pyrroloquinoline quinone-dependent enzymatic bioanode: incorporation of the substituted polyaniline conducting polymer as a mediator. ACS Catal 4(7):2241–2248. doi:10.1021/cs500442b
14. Willner I, Katz E, Lapidot N, Bauerle P (1992) Bioelectrocatalysed reduction of nitrate utilizing polythiophene bipyridinium enzyme electrodes. Bioelectrochem Bioenerg 29(1):29–45
15. Calvo E, Etchenique R, Danilowicz C, Diaz L (1996) Electrical communication between electrodes and enzymes mediated by redox hydrogels. Anal Chem 68(23):4186–4193
16. Koide S, Yokoyama K (1999) Electrochemical characterization of an enzyme electrode based on a ferrocene-containing redox polymer. J Electroanal Chem 468(2):193–201
17. Meredith MT, Kao D-Y, Hickey D, Schmidtke DW, Glatzhofer DT (2011) High current density ferrocene-modified linear poly(ethylenimine) bioanodes and their use in biofuel cells. J Electrochem Soc 158(2):B166–B174. doi:10.1149/1.3505950
18. Hickey DP, Halmes AJ, Schmidtke DW, Glatzhofer DT (2014) Electrochemical characterization of glucose bioanodes based on tetramethylferrocene-modified linear poly(ethylenimine). Electrochim Acta 149:252–257
19. Degani Y, Heller A (1989) Electrical communication between redox centers of glucose oxidase and electrodes via electrostatically and covalently bound redox polymers. J Am Chem Soc 111(6):2357–2358
20. Zakeeruddin S, Fraser D, Nazeeruddin MK, Grätzel M (1992) Towards mediator design: characterization of tris-(4, 4'-substituted-2, 2'-bipyridine) complexes of iron (II), ruthenium (II) and osmium (II) as mediators for glucose oxidase of Aspergillus niger and other redox proteins. J Electroanal Chem 337(1):253–283
21. Nakabayashi Y, Omayu A, Yagi S, Nakamura K, Motonaka J (2001) Evaluation of osmium (II) complexes as electron transfer mediators accessible for amperometric glucose sensors. Anal Sci 17(8):945–950
22. Gallaway JW, Barton SAC (2008) Kinetics of redox polymer-mediated enzyme electrodes. J Am Chem Soc 130(26):8527–8536. doi:10.1021/ja0781543
23. Daigle F, Leech D (1997) Reagentless tyrosinase enzyme electrodes: effects of enzyme loading, electrolyte pH, ionic strength, and temperature. Anal Chem 69(20):4108–4112
24. Kenausis G, Chen Q, Heller A (1997) Electrochemical glucose and lactate sensors based on "wired" thermostable soybean peroxidase operating continuously and stably at 37 C. Anal Chem 69(6):1054–1060

25. Mao L, Yamamoto K (2000) Amperometric on-line sensor for continuous measurement of hypoxanthine based on osmium-polyvinylpyridine gel polymer and xanthine oxidase bienzyme modified glassy carbon electrode. Anal Chim Acta 415(1):143–150

26. Gaspar S, Habermüller K, Csöregi E, Schuhmann W (2001) Hydrogen peroxide sensitive biosensor based on plant peroxidases entrapped in Os-modified polypyrrole films. Sensors Actuators B Chem 72(1):63–68

27. Antiochia R, Gorton L (2007) Development of a carbon nanotube paste electrode osmium polymer-mediated biosensor for determination of glucose in alcoholic beverages. Biosens Bioelectron 22(11):2611–2617

28. Rasmussen M, West R, Burgess J, Lee I, Scherson D (2011) Bifunctional trehalose anode incorporating two covalently linked enzymes acting in series. Anal Chem 83(19):7408–7411. doi:10.1021/ac2014417

29. Mano N, Mao F, Heller A (2003) Characteristics of a miniature compartment-less glucose-O2 biofuel cell and its operation in a living plant. J Am Chem Soc 125(21):6588–6594

30. Rasmussen M, Ritzmann RE, Lee I, Pollack AJ, Scherson D (2012) An implantable biofuel cell for a live insect. J Am Chem Soc 134(3):1458–1460

31. Ohara TJ, Rajagopalan R, Heller A (1994) "Wired" enzyme electrodes for amperometric determination of glucose or lactate in the presence of interfering substances. Anal Chem 66(15):2451–2457

32. Forster RJ, Vos JG (1990) Synthesis, characterization, and properties of a series of osmium- and ruthenium-containing metallopolymers. Macromolecules 23(20):4372–4377

33. Lay PA, Sargeson AM, Taube H, Chou MH, Creutz C (1986) Cis-Bis (2, 2′-Bipyridine-N, N′) Complexes of Ruthenium (III)/(II) and Osmium (III)/(II). Inorg Synth 24:291–299

34. Milton RD, Giroud F, Thumser AE, Minteer SD, Slade RC (2013) Hydrogen peroxide produced by glucose oxidase affects the performance of laccase cathodes in glucose/oxygen fuel cells: FAD-dependent glucose dehydrogenase as a replacement. Phys Chem Chem Phys 15(44):19371–19379

35. Gao F, Courjean O, Mano N (2009) An improved glucose/O2 membrane-less biofuel cell through glucose oxidase purification. Biosens Bioelectron 25:356–361

Chapter 14

Ferrocene-Modified Linear Poly(ethylenimine) for Enzymatic Immobilization and Electron Mediation

David P. Hickey

Abstract

Enzymatic glucose biosensors and biofuel cells make use of the electrochemical transduction between an oxidoreductase enzyme, such as glucose oxidase (GOx), and an electrode to either quantify the amount of glucose in a solution or generate electrical energy. However, many enzymes including GOx are not able to electrochemically interact with an electrode surface directly, but require an external electrochemical relay to shuttle electrons to the electrode. Ferrocene-modified linear poly(ethylenimine) (Fc-LPEI) redox polymers have been designed to simultaneously immobilize glucose oxidase (GOx) at an electrode and mediate electron transfer from their flavin adenine dinucleotide (FAD) active site to the electrode surface. Cross-linked films of Fc-LPEI create hydrogel networks that allow for rapid transport of glucose, while the covalently bound ferrocene moieties are able to facilitate rapid electron transfer due to the ability of ferrocene to exchange electrons between adjacent ferrocene residues. For these reasons, Fc-LPEI films have been widely used in the development of high current density bioanode materials. This chapter describes the synthesis of a commonly used dimethylferrocene-modified linear poly(ethylenimine), as well as the subsequent preparation and electrochemical characterization of a GOx bioanode film utilizing the synthesized polymer.

Key words Dimethylferrocene, Bioanode, Glucose oxidase, Hydrogel, Mediator, Glucose biosensor

1 Introduction

Enzymatic bioelectrode materials have been of great interest for their use in electrochemical biosensors and biological fuel cells. In such materials, oxidoreductase enzymes capable of selectively oxidizing a desired substrate, are immobilized at the surface of an electrode to allow for efficient transduction of an electrochemical signal upon reaction with the molecule of interest. The most widely studied oxidoreductase, glucose oxidase (GOx), is a large glycoprotein used for the commercial detection of glucose and contains a non-dissociable flavin adenine dinucleotide (FAD) cofactor located at the center of the enzyme's active site. The inaccessibility of the FAD site of GOx prevents efficient electrochemical communication with an electrode surface, and therefore, GOx is unable to

Shelley D. Minteer (ed.), *Enzyme Stabilization and Immobilization: Methods and Protocols*, Methods in Molecular Biology, vol. 1504, DOI 10.1007/978-1-4939-6499-4_14, © Springer Science+Business Media New York 2017

participate in direct electron transfer. This problem was addressed in the 1980s and 1990s by several research group who showed that GOx can be electrochemically "wired" to an electrode through the use of redox mediators as electron relays that can nonspecifically bind to the enzyme near the FAD active site and funnel electrons from GOx to the electrode surface [1–3]. While there are several nuanced factors that determine a redox species' ability to behave as an efficient mediator for a given enzyme, all electrochemical mediators must undergo relatively rapid rates of electron self-exchange (the rate at which an electron can be transferred between two identical redox species) and must exhibit a redox potential to allow for thermodynamically favorable electron transfer (i.e., it must induce an electrochemical overpotential). One of the most versatile and widely studied compounds used to mediate electron transfer with GOx is an organometallic iron complex, ferrocene.

Ferrocene (Fc) is a sandwich complex consisting of an iron atom surrounded by two anionic cyclopentadiene (Cp) rings that are sufficiently electron rich to allow the central Fe atom to be stable in either a 2+ or 3+ oxidation state resulting in an overall neutral or cationic species. The chemical stability of both the oxidized (Fc^+) and reduced (Fc) forms of ferrocene, together with its relatively small hydrodynamic Stokes radius (~4 Å) allow it to undergo exceptionally rapid rates of electron self-exchange [4, 5]. Additionally, the overlap between the p-orbitals of the Cp rings and the d-orbitals of the Fe center allow the electrochemical potential ($E_{1/2}$) to be altered through covalent modification of the Cp rings. Therefore, the $E_{1/2}$ of ferrocene can range from –0.17 V for decamethylferrocene to 0.77 V for cyanoferrocene (potentials vs. SCE) [6]. Recent research has made use of ferrocene-modified polymer hydrogels to simultaneously immobilize enzymes as well as provide a proximal redox mediator.

Glatzhofer and Schmidtke *et al.* have developed the use of cross-linked linear poly(ethylenimine) hydrogel films that have been covalently modified with various methyl-, propyl-, and hexylferrocene pendant chains to act as an electron mediation scaffold for a wide range of FAD- and PQQ-dependent enzymes including wild type and mutant GOx, fructose dehydrogenase, lactate oxidase, and alcohol dehydrogenase [7–12]. Additionally, polymethylated ferrocene derivatives have been established as high current density electron mediators, while providing a significantly lower overpotential.

Herein, we provide a detailed procedure for the synthetic preparation of ferrocene-modified LPEI and for subsequently cross-linking it in the presence of GOx in the preparation of an enzymatic bioelectrode material. It should be noted that the following is a procedure for the preparation of dimethylferrocene-modified LPEI; however, the same procedure can be used as a template method for preparing ferrocene-modified LPEI with up to approximately 50% of the polymer backbone substituted.

This can be accomplished by starting with either the commercially available ferrocene or tetramethylferrocenes, which can be prepared using previously reported synthetic procedures [5].

2   Materials

Prepare all aqueous solutions using ultrapure water (obtained by purification of deionized water to attain a sensitivity of 18 MΩ cm at 25 °C) and reagent or analytical grade reagents and reactants.

2.1   Synthesis of Dimethylferrocene-Modified Linear Poly(ethylenimine)

2.1.1   Linear Poly(ethylenimine) Synthesis Preparations

1. Hydrochloric acid (HCl) reagent grade, store in a ventilated cabinet at room temperature.

2. 3 M HCl solution: Prepare a 2 L stock solution of 3 M HCl by slowly adding 500 mL of 12 M HCl to 1.5 L of water.

3. Linear poly(ethyloxazoline) (MW 200,000, Sigma).

4. Sodium hydroxide pellets (NaOH, Sigma), store under dry ambient conditions.

2.1.2   3-Bromopropionyl Ferrocene Synthesis Preparations

1. Methylene chloride (CH_2Cl_2), store in a ventilated cabinet at room temperature.

2. Anhydrous aluminum chloride ($AlCl_3$) powder, store under dry conditions in the presence of a desiccant.

3. 3-Bromopropionyl chloride (Sigma) technical grade, store at 2–8 °C.

4. Aluminum chloride–3-bromopropionyl chloride complex: Add 5 mL of CH_2Cl_2 to 0.223 g (1.67 mmol) of aluminum chloride ($AlCl_3$) in a 25 mL round bottom flask with a stir bar to form a suspension of $AlCl_3$. Add 0.286 g (1.67 mmol) of 3-bromopropionyl chloride to the stirring $AlCl_3$ suspension dropwise at 0 °C. Let the mixture warm to room temperature and stir until all of the $AlCl_3$ is dissolved (~1 h after the addition of 3-bromopropionyl chloride).

5. Saturated sodium bicarbonate solution: Prepare a 1 L stock solution of saturated sodium bicarbonate ($NaHCO_3$) by dissolving 9.6 g of $NaHCO_3$ into 1 L of water.

6. Brine solution: Prepare a 1 L stock solution of brine by dissolving 35.7 g of sodium chloride (NaCl) into 1 L of water.

7. Anhydrous magnesium sulfate ($MgSO_4$), store under dry ambient conditions.

2.1.3   3-Bromopropyl Ferrocene Synthesis Preparations

1. Methylene chloride (CH_2Cl_2), store in a ventilated cabinet at room temperature.

2. Borane *tert*-butylamine complex (Sigma), store under anhydrous conditions in the presence of a desiccant.

3. Anhydrous aluminum chloride ($AlCl_3$) powder, store under dry conditions in the presence of a desiccant.

4. Aluminum chloride–*tert*-butylamine borane complex solution: Add 0.278 g (2.09 mmol) $AlCl_3$ into a small (≤ 25 mL) round bottom flask with 5 mL methylene chloride. Dissolve 0.364 g (4.19 mmol) of *tert*-butylamine borane complex (with respect to the ferrocene product from the previous reaction) in 2 mL methylene chloride. Add the reducing solution dropwise to the $AlCl_3$ mixture at 0 °C. Stir the mixture at room temperature under N_2 for 1 h.

5. Anhydrous magnesium sulfate ($MgSO_4$), store under dry ambient conditions.

6. Silica powder.

7. Acetonitrile, commercially available, reagent grade, store in a ventilated cabinet at room temperature.

2.1.4   Synthesis of Ferrocene-Modified Linear Poly(ethylenimine)

1. Acetonitrile, commercially available, reagent grade, store in a ventilated cabinet at room temperature.

2. Methanol, commercially available, reagent grade, store in a ventilated cabinet at room temperature.

2.2   Dimethylferrocene-Mediated Glucose Oxidase Bioanodes

2.2.1   Preparation of Dimethylferrocene-LPEI/ Glucose Oxidase Electrode Films

1. Glucose oxidase from *Aspergillus niger* (Type X-S, lyophilized powder, 100,000–250,000 U/g solid, Sigma), store at −20 °C.

2. Ethylene glycol diglycidyl ether (EGDGE), commercially available, store under anhydrous conditions at room temperature.

3. Glassy carbon electrodes (3 mm disk electrodes), polished immediately prior to use.

4. Ferrocene-modified LPEI solution: Add 200 µL of water to 4 mg of $FcMe_2$-LPEI in a small 1 dram vial. Add 15 µL aliquots of 0.1 M HCl (the total addition of HCl should NOT exceed 200 µL) until the polymer is dissolved. Add enough water to bring the total solution volume to 399 µL. Use a small (less than 1 µL) amount of concentrated HCl to adjust the pH of the polymer solution to pH 5.0 ± 0.2. The final polymer concentration should be 10 mg mL^{-1}.

5. Glucose oxidase (GOx) solution: Dissolve 1 mg of GOx using 100 µL of water. Mix the solution rapidly using a vortex generator for 15 s. The final enzyme concentration should be 10 mg mL^{-1}.

6. Cross-linker solution: Add 5 µL of ethylene glycol diglycidyl-ether (EGDGE) to 45 µL of water. Mix the solution rapidly using a vortex generator for 15 s [Note: prepare this solution shortly before use, to prevent degradation of the cross-linker through hydrolysis].

1. Phosphate buffer: 100 mM and pH 7.4.

2. 2 M β-D-glucose aqueous solution, store at 4 °C (*see* **Note 1**).

---

3 Methods

3.1 Dimethylferro cene-Modified Linear Poly(ethylenimine) Synthesis

Linear poly(ethylenimine) (LPEI) (MW approximately 80,000) is prepared through hydrolysis of linear poly(2-ethyl-2-oxazoline) (MW 200,000) from Sigma Aldrich [13]. The following procedure should allow for the preparation of ~20 g of LPEI.

3.1.1 Synthesis of Linear Poly(ethylenimine)

1. Dissolve 30.0 g of linear poly(2-ethyl-2-oxazoline) in 1.80 L of aqueous 3 M HCl in a 3 L round bottom flask.

2. Stir the reaction solution for 5 days at 110 °C, using a jacketed reflux condenser to prevent excessive evaporation of solvent.

3. Allow the solution to cool to room temperature, then evaporate the solvent under reduced pressure (*see* **Note 2**).

4. Dissolve the crude polymer in 3 L of water, then add solid NaOH until the pH of the solution is greater than pH 10.0. Neutralization causes the polymer to crash out of solution, and thus the neutralization may not occur uniformly. To ensure that the polymer is neutralized, heat the solution gently until the polymer dissolves, then allow it to cool to room temperature (*see* **Note 3**).

5. Collect the crystallized polymer in a fritted glass filter, using vacuum filtration to speed the filtration process.

6. Redissolve the crystallized polymer in 3 L of warm water, then allow the solution to cool to room temperature (thus reprecipitating the polymer) and again collect the polymer precipitate by filtration through a fritted glass filter. Repeat this process until the pH of the filtrate is neutral.

7. Collect the solid neutralized polymer in a small glass jar and dry in a vacuum oven at ~50 °C for 1 day then for ~70 °C for 1 day (*see* **Note 4**).

8. Confirm polymer structure by ^{1}H-NMR in CD$_3$OD: δ(ppm) 2.65 (4H, broad *t*, -CH$_2$-N).

3.1.2 Synthesis of 3-Bromopropionyl Dimethylferrocene

1. Dissolve 0.30 g (1.40 mmol) dimethylferrocene into a 50 mL round bottom flask in 15 mL methylene chloride with a stir bar.

2. Add the previously prepared solution of AlCl$_3$*3-bromopropionyl chloride complex dropwise to the stirring dimethylferrocene solution at 0 °C (*see* **Notes 5** and **6**).

3. Cover the solution and stir the reaction mixture at room temperature for 12–16 h.

4. Dilute the reaction mixture with 20 mL methylene chloride and quench the reaction by pouring the solution over an equivalent volume of ice in an Erlenmeyer flask.

5. Stir the mixture until all of the ice is melted (~30 min).

6. Using a separatory funnel, separate the organic portion and wash it with 50 mL of saturated sodium bicarbonate ($NaHCO_3$) followed by 50 mL of brine solution.

7. Filter the organic portion through a plug of magnesium sulfate ($MgSO_4$) in a coarse fritted glass filter. Wash the $MgSO_4$ with 25 mL CH_2Cl_2 to remove any excess product.

8. Transfer the dissolved product into a round bottom flask, and reduce the solvent to approximately 25 mL under reduced pressure at 40 °C (*see* **Note 7**).

3.1.3 Synthesis of 3-Bromopropyl Dimethylferrocene

1. Add the propionyl dimethylferrocene solution from Subheading 3.2 dropwise into the stirring premade solution of *tert*-butylamine-borane/$AlCl_3$ complex at 0 °C.

2. Allow the mixture to warm to room temperature and stir for overnight (16–24 h) at ambient temperature. Cover the reaction vessel to prevent evaporation of the solvent.

3. Quench the reaction by adding 50 mL DI water slowly to the reaction mixture and stirring for 1 h (*see* **Note 8**).

3.1.4 Purification of 3-Bromopropyl Dimethylferrocene

1. Separate the organic portion using a separatory funnel, and wash it with 50 mL of saturated $NaHCO_3$.

2. Remove excess water from the organic fraction by passing it through a plug of magnesium sulfate ($MgSO_4$) in a fritted glass filter, and remove the organic solvent by rotatory evaporation.

3. The crude product should be a red–orange oil containing solid crystalline particles (these particles are excess *tert*-butylamine borane). Dissolve product in 5–10 mL acetonitrile and pass through a filter of flash silica to remove excess *tert*-butylamine borane. Rinse the filter with excess acetonitrile to ensure that all of the product is collected (*see* **Note 9**).

4. Collect the filtered product in a 25 mL round bottom flask and evaporate the excess acetonitrile under reduced pressure. The yield over two steps from dimethylferrocene is around 65 %.

5. Confirm the structure of the product by ^1H-NMR in $CDCl_3$ (*see* Fig. 1): δ(ppm) 1.91 (3H, s, Fc(–CH_3)), 1.96 (3H, s, Fc(–CH_3)), 2.01 (2H, m, C-CH_2–C), 2.44 (2H, t, Fc–CH_2–C), 3.41 (2H, t, C–CH_2–Br), 3.92 (7H, m, Fc–H).

3.1.5 Synthesis of Ferrocene-Modified Linear Poly(ethylenimine)

1. Add 0.100 g (2.33 mmol) of LPEI to a small round bottom flask with 7 mL acetonitrile (*see* **Note 10**).

2. Stir the LPEI mixture and heat to 80 °C with a reflux condenser, then add methanol dropwise until all of the LPEI dissolves (≤2 mL).

a

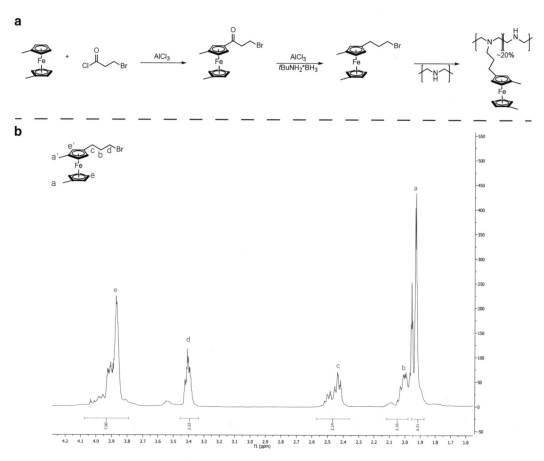

b

Fig. 1 (**a**) Synthetic reaction scheme for the preparation of dimethylferrocene-modified linear poly(ethylenimine), (**b**) ¹H-NMR of 3-bromopropyl dimethylferrocene in CDCl₃

3. Dissolve 0.156 g (0.47 mmol) 3-bromopropyl dimethylferrocene in 2 mL acetonitrile, and add to the stirring LPEI solution (*see* **Note 11**).

4. Stir the reaction mixture 24 h at 90–100 °C using a jacketed condenser to prevent excessive solvent evaporation.

5. Cool the reaction solution to room temperature, and evaporate the solvent under reduced pressure.

6. Add 20 mL of diethyl ether to the reaction flask and allow the final product to soak for 1 h to remove any unreacted bromopropyl dimethylferrocene.

7. Decant the diethyl ether and evaporate excess solvent under reduced pressure to give the dimethylferrocene-modified LPEI (FcMe₂-LPEI) product.

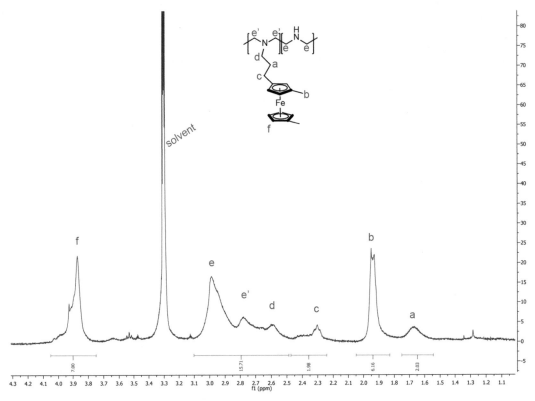

Fig. 2 ¹H-NMR of dimethylferrocene-modified linear poly(ethylenimine) (FcMe₂-LPEI) in CD₃OD

3.1.6 Characterization of Ferrocene-Modified Linear Poly(ethylenimine)

1. Confirm structure and substitution of the product by ¹H-NMR in CD₃OD (*see* Fig. 2): δ(ppm) 1.67 (2H, C–CH₂–C), 1.94 (6H, Fc(–CH₃)), 2.30 (2H, Fc-CH2–C), 2.50–3.10 (16H, N–CH₂–C, polymer backbone), 3.87 (7H, Fc-H).

3.2 Dimethylferrocene-Mediated Glucose Oxidase Bioanode

3.2.1 Preparation of Dimethylferrocene-LPEI/ Glucose Oxidase Electrode Films

1. Combine 14 μL of FcMe₂-LPEI solution (10 mg/mL), 6 μL of GOx solution (10 mg/mL) and 0.75 μL of cross-linker solution (10 vol/vol.% EGDGE) in a small centrifuge vial.

2. Vortex vigorously for 20 s, then coat 3 mL aliquots of the solution onto the surface of each of three glassy carbon electrodes using an Eppendorf pipette (*see* **Note 12**).

3. Allow electrode films to cure and dry for 24 h at room temperature under ambient atmosphere.

3.2.2 Characterization of Dimethylferrocene-LPEI/ Glucose Oxidase Electrode Films

1. Dimethylferrocene-LPEI/GOx films are characterized electrochemically in a 100 mM phosphate buffer, pH 7.4, at 25 °C using a standard three electrode configuration with the FcMe₂-LPEI/ GOx glassy carbon acting as the working electrode, a saturated calomel (SCE) reference electrode and a Pt counter electrode.

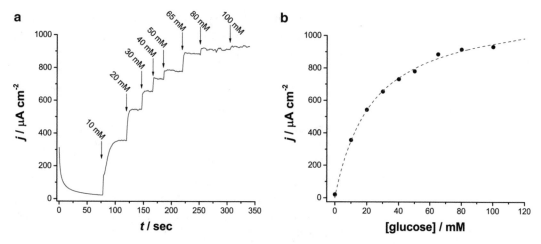

Fig. 3 (**a**) Constant potential amperometric *i* vs. *t* curve of FcMe$_2$-LPEI/GOx film response to injections of glucose while being held at 0.5 V vs SCE, and (**b**) the resulting Michaelis–Menten plot. Experiments were performed using 100 mM phosphate buffer at 25 °C

2. Perform preliminary cyclic voltammetry (CV) to allow for the polymer films to electrochemically equilibrate; 10 scans, –0.1 V to 0.6 V vs. SCE at 50 mV s^{-1}.

3. Test the electrode kinetics by performing slow scan rate CVs in the absence and presence of various concentrations of substrate (glucose); four scans, –0.1 V to 0.6 V vs. SCE at 2 mV s^{-1}, in the presence of 0 mM glucose and 100 mM glucose.

4. The apparent enzymatic electrode kinetics are commonly determined using constant potential amperometry in which an oxidizing potential is held and the current is monitored as a response to the addition of substrate (glucose) to a stirring solution. Hold the potential at 0.1 V more positive than the peak oxidation potential ($E_{p, a}$); allow the background current to equilibrate to a constant value (typically ~100 s), then inject aliquots of 2 M glucose solution so that the total glucose concentration is 10, 20, 30, 40, 50, 65, 70, and 100 mM. Effective FcMe$_2$-LPEI/GOx films should have a maximum current density around 1.0 mA cm^{-2} (*see* Fig. 3).

4   Notes

1. Prepare at least 24 h prior to use.

2. Alternatively the solvent can be removed by vacuum distillation.

3. Neutralization of the polymer solution requires a large quantity of NaOH (>10 g) and is highly exothermic; caution should be exercised during this step.

4. Shortly after removing the polymer from the oven, use a razor to scrape the polymer from the sides of the jar. This will make it easier to handle the polymer once it cools.

5. It is very important to add the $AlCl_3$*3-bromopropionyl chloride complex dropwise; rapid addition of the complex can result in diacylation of dimethylferrocene.

6. There should be a very noticeable color change of the ferrocene solution upon addition of the $AlCl_3$*3-bromopropionyl chloride complex from orange to purple.

7. Do NOT remove all of the solvent; the crude product becomes slightly unstable once isolated. The highest overall yields are obtained when the dissolved product is used in subsequent steps without further purification.

8. The reaction between excess borane in solution and water is exothermic and produces hydrogen gas; use caution during this step.

9. The unreacted *tert*-butylamine borane is only partially soluble in acetonitrile; a small amount of CH_2Cl_2 can be used to help load the crude product onto the filter.

10. LPEI will NOT dissolve immediately at 25 °C, heat must be applied first.

11. For best results, wash the flask with an additional 1 mL acetonitrile to collect any remaining bromopropyl dimethylferrocene.

12. The coating process is a mixture of drop-coating and paint coating, in which 3 µL of the film solution is dispensed onto the electrode and the tip of the pipette is used to evenly distribute the solution across the electrode surface.

References

1. Degani Y, Heller A (1987) Direct electrical communication between chemically modified enzymes and metal electrodes. I. Electron transfer from glucose oxidase to metal electrodes via electron relays, bound covalently to the enzyme. J Phys Chem 91(6):1285–1289. doi:10.1021/j100290a001

2. Degani Y, Heller A (1989) Electrical communication between redox centers of glucose oxidase and electrodes via electrostatically and covalently bound redox polymers. J Am Chem Soc 111(6):2357–2358. doi:10.1021/ja00188a091

3. Schuhmann W, Ohara TJ, Schmidt HL, Heller A (1991) Electron transfer between glucose oxidase and electrodes via redox mediators bound with flexible chains to the enzyme surface. J Am Chem Soc 113(4):1394–1397. doi:10.1021/ja00004a048

4. Struchkov YT, Andrianov VG, Sal'nikova TN, Lyatifov IR, Materikova RB (1978) Crystal and molecular structures of two polymethylferrocenes: sym-octamethylferrocene and decamethylferrocene. J Organomet Chem 145(2):213–223. doi:10.1016/S0022-328X(00)91127-6

5. Meredith MT, Hickey DP, Redemann JP, Schmidtke DW, Glatzhofer DT (2013) Effects of ferrocene methylation on ferrocene-modified linear poly(ethylenimine) bioanodes. Electrochim Acta 92:226–235. doi:10.1016/j.electacta.2013.01.006

6. Britton WE, Kashyap R, El-Hashash M, El-Kady M, Herberhold M (1986) The

anomalous electrochemistry of the ferrocenylamines. Organometallics 5(5):1029–1031. doi:10.1021/om00136a033

7. Merchant SA, Tran TO, Meredith MT, Cline TC, Glatzhofer DT, Schmidtke DW (2009) High-sensitivity amperometric biosensors based on ferrocene-modified linear poly(ethylenimine). Langmuir 25(13):7736–7742. doi:10.1021/la9004938

8. Meredith MT, Kao D-Y, Hickey D, Schmidtke DW, Glatzhofer DT (2011) High current density ferrocene-modified linear poly(ethylenimine) bioanodes and their use in biofuel cells. J Electrochem Soc 158(2):B166–B174. doi:10.1149/1.3505950

9. Hickey DP, Halmes AJ, Schmidtke DW, Glatzhofer DT (2014) Electrochemical characterization of glucose bioanodes based on tetramethylferrocene-modified linear poly(ethylenimine). Electrochim Acta 149:252–257. doi:10.1016/j.electacta.2014.10.077

10. Hickey DP, Giroud F, Schmidtke DW, Glatzhofer DT, Minteer SD (2013) Enzyme cascade for catalyzing sucrose oxidation in a biofuel cell. ACS Cat 3(12):2729–2737. doi:10.1021/cs4003832

11. Milton RD, Wu F, Lim K, Abdellaoui S, Hickey DP, Minteer SD (2015) Promiscuous glucose oxidase: electrical energy conversion of multiple polysaccharides spanning starch and dairy milk. ACS Cat 5(12):7218–7225. doi:10.1021/acscatal.5b01777

12. Hickey DP, Reid RC, Milton RD, Minteer SD (2016) I. Biosens Bioelectron 77:26–31. doi:10.1016/j.bios.2015.09.013

13. York S, Frech R, Snow A, Glatzhofer D (2001) A comparative vibrational spectroscopic study of lithium triflate and sodium triflate in linear poly(ethylenimine). Electrochim Acta 46(10–11):1533–1537. doi:10.1016/S0013-4686(00)00749-0

Chapter 15

FAD-Dependent Glucose Dehydrogenase Immobilization and Mediation Within a Naphthoquinone Redox Polymer

Ross D. Milton

Abstract

Electrochemically-active polymers (redox polymers) are useful tools for simultaneous immobilization and electron transfer of enzymes at electrode surfaces, which also serve to increase the localized concentration of the biocatalyst. The properties of the employed redox couple must be compatible with the target biocatalyst from both an electrochemical (potential) and biochemical standpoint. This chapter details the synthesis of a naphthoquinone-functionalized redox polymer (NQ-LPEI) that is used to immobilize and electronically communicate with flavin adenine dinucleotide-dependent glucose dehydrogenase (FAD-GDH), yielding an enzymatic bioanode that is able to deliver large catalytic current densities for glucose oxidation at a relatively low associated potential.

Key words Naphthoquinone, Redox polymer, Glucose dehydrogenase, Enzymatic fuel cell, Glucose oxidation, Enzymes

1 Introduction

Enzymatic fuel cells (EFCs) are devices that employ biological catalysts, such as enzymes, to facilitate the oxidation of alternative energy-dense fuels (i.e., sugars) under mild conditions (such as neutral pH and room temperature) in comparison to typical H_2/O_2 fuel cells [1–3]. Additionally, EFCs typically incorporate an enzyme at the biocathode of devices (usually a multi-copper oxidase) that can accommodate the 4-electron reduction of O_2 to H_2O [4]. It is favorable to immobilize enzymes at electrode surfaces, whereby an enhanced localized concentration of the biocatalyst offers improved catalytic current densities and lower overall costs of devices; spent fuel can be easily replenished without the requirement to replace a diffusive enzyme.

The mechanism of electron transfer between an enzyme and an electrode surface is of extreme importance (Fig. 1). Mediated electron transfer (MET) utilizes a small redox-active species to shuttle electrons between an enzyme's redox-active cofactor and an

Shelley D. Minteer (ed.), *Enzyme Stabilization and Immobilization: Methods and Protocols*, Methods in Molecular Biology, vol. 1504, DOI 10.1007/978-1-4939-6499-4_15, © Springer Science+Business Media New York 2017

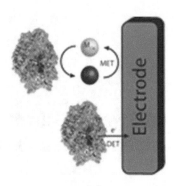

Fig. 1 Mediated electron transfer (*top*) and direct electron transfer (*bottom*) between an enzyme and an electrode surface

electrode surface. In contrast, direct electron transfer (DET) takes place when the redox-active cofactor of an enzyme is proximally located such that the distance between the cofactor and the electrode is small enough to enable electron tunneling to or from the electrode surface. Additionally, DET must not block access of the substrate to the enzyme [2]. Redox polymers simultaneously provide a support to immobilize enzymes at an electrode surface while also locally providing redox active electron mediators that are immobilized and do not diffuse into the bulk solution. Further, the ability of redox species to undergo "self-exchange" affords larger 3D structures of immobilized enzyme-redox polymer architectures, which in turn enables larger catalytic current densities. While DET has been firmly established for many multi-copper oxidases (such as laccase and bilirubin oxidase), MET firmly remains the principal mechanism for electron transfer at the bioanode of EFCs.

When considering the use of an electron mediator, the potential of the redox species plays a major role in the mediator's compatibility. For an oxidative enzymatic reaction, the electron mediator must possess a redox potential that is more positive than the redox potential of the enzyme's redox-active cofactor. An increased potential difference between these two components typically offers improved overall catalytic turnover (with a minimal potential difference being required for optimal catalytic turnover). In terms of improving an EFC's performance, a maximized potential difference between the anode and cathode of an EFC would yield improved power densities; similarly, improved catalytic currents at the anode and cathode would *also* yield improve power densities. Unfortunately, these two variables are mutually exclusive.

The redox potential alone of the redox species does not constitute a "good" electron mediator. The redox species must be able to accept electrons from the cofactor of the enzyme and enzymatic rejection of the redox species from the active site of the enzyme (i.e., the mediator is a poor "substrate") may prevent electron transfer (depending on the mechanism of electron transfer between

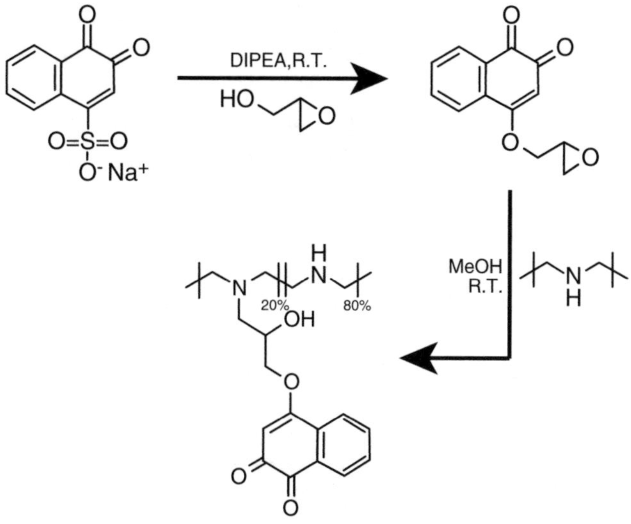

Fig. 2 Synthetic route for the preparation of naphthoquinone-functionalized linear poly(ethylenimine) (NQ-LPEI)

the redox species and the cofactor). An example of this is found in this chapter; the naphthoquinone redox species and its associated redox polymer is able to efficiently transfer electrons between flavin adenine dinucleotide-dependent glucose dehydrogenase (FAD-GDH) although its counterpart, glucose oxidase (GOx), employs the same cofactor (with comparable properties) yet is unable to undergo mediated bioelectrocatalytic glucose oxidation [5].

The naphthoquinone redox polymer (synthesized as per Fig. 2 and detailed below) possesses a relatively low redox potential and is able to undergo MET with fungal FAD-GDH resulting in large catalytic currents for mediated bioelectrocatalytic glucose oxidation.

2 Materials

All aqueous solutions should be prepared using ultrapure water (18 MΩ cm at 25 °C).

2.1 Reagents and Solvents for the Synthesis and Purification of 1,2-Naphthoquinone-4-glycidol

1. 1,2-Naphthoquinone-4-sulfonic acid sodium salt (NQS): used as received.

2. Diisopropylethylamine (DIPEA): used as received (*see* **Note 1**).

3. (±)-glycidol: stored at 4 °C and used as received.

4. Brine/saturated sodium chloride solution: add approximately 90 g of sodium chloride (reagent grade) to 250 mL of distilled water. Shake the solution with the aid of sonication to dissolve the sodium chloride. Insoluble sodium chloride should remain.

5. Liquid–liquid extraction solutions: (A) 100 mL of a brine solution is added to 35 mL of dichloromethane. (B) 100 mL of distilled water is added to 35 mL of chloroform.

6. Magnesium sulfate: anhydrous reagent is added loaded on top of a fritted glass filter to form a layer of approximately 1 cm thickness. Solutions are dried over the magnesium sulfate layer by gravity.

7. Silica gel purification solvents: Diethyl ether is mixed with dichloromethane at a 3:1 ratio.

2.2 Bioelectrode Components and Electrochemistry Solutions

1. Enzyme solution: FAD-dependent glucose dehydrogenase (fungal from *Aspergillus* sp.) is suspended in ultrapure water at a concentration of 30 mg/mL. Typically, 1–2 mg of lyophilized FAD-dependent glucose dehydrogenase is used. Prepared immediately prior to use (*see* **Note 2**).

2. Cross-linker solution: Ethylene glycol diglycidyl ether (EGDGE) is diluted with ultrapure water to a concentration of 10% v/v. Prepared immediately prior to use (*see* **Note 3**).

3. Multi-walled carbon nanotube suspension: COOH-functionalized MWCNTs are added to isopropanol at a final concentration of 5 mg/mL. The resulting suspension is disrupted by sonication for 1 h. The final suspension is then left to stand for an additional hour prior to use. Store at room temperature (*see* **Note 4**).

4. Toray carbon paper electrodes: Untreated Toray carbon paper is cut into a flag shape with a 1 cm^2 "flag". The remaining electrode is coated with melted paraffin wax (*see* **Note 5**).

5. Electrochemistry buffers and solutions: A citrate/phosphate buffer (0.2 M) is prepared and the pH is adjusted to 6.5. A glucose solution (1 M) is prepared using the citrate/phosphate buffer solution at least 24 h prior to electrochemical evaluation (to allow for mutarotation of glucose). Store at 4 °C.

6. Electrochemistry equipment: Bioelectrodes are initially evaluated using cyclic voltammetry with a potentiostat operating in a standard 3-electrode half-cell configuration. Typically, a large platinum counter electrode is used along with common reference electrode (such as the saturated calomel electrode, Ag/AgCl electrode or standard hydrogen electrode).

2.3 Linear Poly(Ethylenimine)

Linear poly(ethylenimine) (LPEI) was prepared as described in Chapter 14 Subheading 3.1.1.

3 Methods

All procedures (with the exception of solvent evaporation) are conducted at room temperature.

3.1 Synthesis of 1,2-Naphthoquinone-4-glycidol

1. Add 241 µL of DIPEA (1.2 mol. eq., 178.8 mg, 1.38 mmol) to a 10 mL stirred solution of excess glycidol (9 g, 121 mmol) in a dropwise fashion. Leave the glycidol solution to stir for 10 min at room temperature (*see* **Note 6**).

2. Add 300 mg of solid 1,2-naphthoquinone-4-sulfonic acid sodium salt (1 mol. eq., 1.15 mmol) to the stirred solution of glycidol in a single portion. Leave the suspension to stir for an additional 10 min (*see* **Note 7**).

3. Transfer the suspension to a separatory funnel that contains the liquid-liquid extraction solution (A). Repeatedly extract the organic fractions with dichloromethane against brine (*see* **Note 8**).

4. Combine all of the organic fractions and remove the solvent *in vacuo* (*see* **Note 9**).

5. The yellow product is dissolved in chloroform and extracted a further three times with the liquid-liquid extraction solution (B). Combine the organic fractions, dry over magnesium sulfate (anhydrous) and remove the solvent *in vacuo* to yield a dry yellow powder (*see* **Note 10**).

3.2 Purification of 1,2-Naphthoquinone-4-glycidol

1. Prepare a column with silica gel for the purification of the 1,2-naphthoquinone-4-glycidol. For a 300 mg preparation, a column of 30 cm in length and 2.5 cm diameter is typically used. A flash column (with a large volume reservoir) is preferred.

2. Equilibrate the silica gel column with 3:1 diethyl ether–dichloromethane. Load the yellow product onto the column (*see* **Note 11**).

3. Purification is afforded by constant flushing with above elution solvent mixture. DIPEA typically elutes from the column at the very beginning of the product band (yellow)—the first ca. 5% of the eluted product (by volume) is usually discarded (*see* **Note 12**).

4. The solvent is removed *in vacuo* to yield a yellow powder. Overall, the entire synthesis and purification typically results in a mol. yield % of approximately 25–30%. The final product is stored in a vacuum desiccator at room temperature, until use.

3.3 Characterization of 1,2-Naphtho quinone-4-glycidol

1. ^{1}H and ^{13}C NMR spectroscopy can be performed in CDCl$_3$ to confirm the successful synthesis of the above product. Additionally, any residual glycidol or DIPEA can be identified. Figure 3a, b presents the ^{1}H and ^{13}C NMR spectra for the product, respectively.

1H NMR (400 MHz, CDCl3): δ 8.06 (d, *J*=7.6 Hz, 1H, -CH), 7.87 (d, *J*=7.8 Hz, 1H, -CH), 7.66 (t, *J*=7.6 Hz, 1H, -CH), 7.55 (t, *J*=7.6 Hz, 1H, -CH), 5.90 (s, 1H, -CH adjacent to ether linkage), 4.45 (dd, *J*=11.3, 2.1 Hz, 1H, -O-*CH2*-), 3.97 (dd, *J*=11.3, 6.4 Hz, 1H, -O-*CH2*-), 3.48–

3.40 (m, 1H, -CH- epoxy), 2.97 (t, *J*=4.4 Hz, 1H, -*CH2*-epoxy), 2.79 (dd, *J*=4.5, 2.5 Hz, 1H, -*CH2*- epoxy).
13C NMR (101 MHz, CDCl3): δ 179.21 (C=O), 179.06 (C=O), 167.18 (C-O), 134.90 (CH), 131.57 (C), 131.53 (CH), 130.19 (C), 128.99 (CH), 124.72 (CH), 103.62 (CH), 70.28 (-O-CH2), 48.87 (epoxy CH), 44.27 (epoxy CH2).
ESI-MS (high resolution) found: 253.0477 [M + Na]+.

3.4 Synthesis of 1,2-Naphthoquinone-4-glycidyl-LPEI Redox Polymer (NQ-LPEI)

The recipe below aims to theoretically prepare a redox polymer with approximately 25% substitution across the polymer backbone/repeating units. Typically, a final substitution of 20–23% is estimated.

1. LPEI (46 mg synthesized as per Chapter 14 Subheading 3.1.1, 4 mol. eq., 1.1 mmol) is dissolved in a stirred solution of methanol (15 mL).

2. To the above, 62 mg of 1,2-naphthoquinone-4-glycidol (1 mol. eq., 0.27 mmol) in a minimal volume of methanol is added dropwise. The resulting solution is left to stir overnight at room temperature.

3. Upon completion, the solvent is removed *in vacuo*. The resulting residue is soaked with two portions of tetrahydrofuran (15 mL, 1 h each) to remove any unsuccessfully grafted 1,2-naphthoquinone-4-glycidol (*see* **Note 13**).

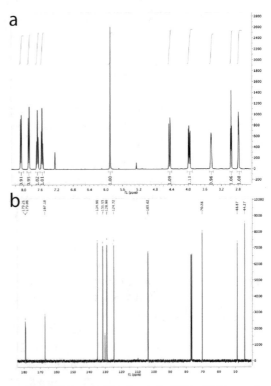

Fig. 3 ¹H (**a**, 400 MHz) and ¹³C (**b**, 100 MHz) NMR spectra of purified 1,2-naphthoquinone-4-glycidol

4. The final polymer (NQ-LPEI) is dried and stored (dry at room temperature) for later use. For the construction of bioelectrodes (below) the resulting polymer is typically dissolved in ultrapure water at a concentration of 10 mg/mL, immediately prior to use (*see* **Note 14**).

3.5 Bioelectrode Construction

The instructions provided below are for the preparation of a bioelectrode whereby the naphthoquinone moiety of the redox polymer facilitates MET between fungal FAD-dependent glucose dehydrogenase and a carbon electrode surface. To date, this configuration has been evaluated on glassy carbon and carbon paper (Toray) electrode surfaces. Ultrapure water (18 MΩ cm at 25 °C) is used for all steps below.

3.5.1 Bioelectrode Construction

1. Using a positive displacement pipette, deposit 5 µL of the MWCNTs suspension onto the surface of a polished glassy carbon electrode to yield an approximate loading of 0.35 mg/cm^2. Proceed to the next step once the electrode is dry (~10 min at room temperature) (*see* **Notes 4** and **15**).

2. Combine the NQ-LPEI polymer solution from Subheading 3.4 with the enzyme solution and the cross-linker solution in a small Eppendorf tube (500 µL) at the following ratio (can be scaled):

   (a) *NQ-LPEI solution*: 84 µL

   (b) *Enzyme solution*: 36 µL

   (c) *Cross-linker solution*: 4.5 µL

   (*see* **Note 16**).

3. Deposit 3 µL of the resulting solution onto the modified surface (MWCNTs) of a glassy carbon electrode and leave to dry overnight at room temperature (*see* **Note 17**).

3.5.2 Bioelectrode Evaluation

The following evaluation procedure is specific to FAD-dependent glucose dehydrogenase. Differing enzymatic properties (such as turnover number) of alternate biocatalysts may require testing under custom conditions (i.e., scan rate).

1. Prior to electrochemical analysis, dried electrodes are gently rinsed with ultrapure water to remove any excess NQ-LPEI.

2. Initially, bioelectrodes are electrochemically cycled for between 10 and 20 complete cycles at 50 mV/s. This further allows for the removal of any unbound NQ-LPEI; during this initial step, the NQ-LPEI film will also hydrate and swell (*see* **Note 18**). The potential window varies depending on the pH of the supporting electrolyte as well as the reversibility of the NQ species at the chosen electrode surface. Typically, the potential window of interest for a modified glassy carbon electrode operating at pH 6.5 is between −0.45 and +0.05 V (vs. a saturated calomel reference electrode).

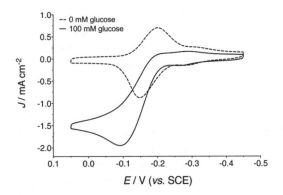

Fig. 4 Representative cyclic voltammogram of a FAD-dependent glucose dehydrogenase bioelectrode prepared using the NQ-LPEI polymer, in the absence (*dashed line*) and presence (*solid line*) of the glucose. Bioelectrodes were analyzed at pH 6.5 (citrate/phosphate buffer, 0.2 M) at a scan rate of 10 mV/s

3. Following this initial equilibration step, the bioelectrodes are then analyzed for a substrate blank (i.e., in the absence of glucose) at 10 mV/s, under the same potential window as described above. Typically, five complete cycles are performed during which time the redox peaks for the NQ species stabilize.

4. Bioelectrodes are then analyzed once again (at 10 mV/s) in the presence of the enzyme's substrate, glucose (Fig. 4).

4   Notes

1. DIPEA is required as a non-nucleophilic sterically hindered base. The use of other amine-containing nucleophilic bases (such as triethylamine) results in a red color change and major undesired side products.

2. Initial findings suggest that FAD-dependent glucose dehydrogenase from *Aspergillus* sp. is favorable. Further subspecies (such as that obtained from *Aspergillus oryzae*) are compatible but diminished catalytic current densities are typically observed.

3. EGDGE is used as a crosslinker between secondary amines on the LPEI backbone and accessible nucleophilic amine acids on the protein surface (and iterations thereof). The polymeric counterpart (PEGDGE) has not been evaluated for this system.

4. Carboxylated MWCNT are purchased from www.cheaptubes.com and used as received.

5. Toray carbon paper (TGP-H-060) is purchased from www.fuelcellearth.com and used as received. Toray carbon paper is ordered without hydrophobic (Teflon) treatment.

6. This step of the reaction can be continued for more than 10 min, although side products (most likely epoxide ring opening-derived) are observed.

7. Over the course of 10 min, the suspension will change from orange to brown in color.

8. To aid in the removal of unreacted glycidol, around 6–8 extractions are typically performed.

9. Extracted organic phases are typically combined and concentrated along the course of extraction. Due to its extraction against brine, dichloromethane has a tendency to "bump" under reduced pressure. Passing the organic extract through a thin bed of anhydrous magnesium sulfate prior to removing the solvent prevents this.

10. This step aids in the removal of residual glycidol. The presence of glycidol in the final product results in an "oily" appearance. A successfully purified product forms a yellow powder when following removal of the chloroform.

11. The product is not overly soluble in the 3:1 diethyl ether–dichloromethane solvent mixture. Typically, the product is dissolved in a *minimal* volume of dichloromethane for application to the column bed (equilibrated with the solvent mixture).

12. Applying the product to the column in a small volume of dichloromethane and the presence of glycidol will result in the mobility of the crude product mixture for ca. 5 % of the column bed length—utilizing a long column bed will aid purification. After the initial mobility has subsided, the product will elute as a yellow solution. Under the above solvent mixture residual 1,2-naphthoquinone-4-sulfonic acid sodium salt remains at the top of the column bed and does not typically elute. The product is eluted with multiple column volumes of the solvent mixture, until the eluent appears almost colorless. Eluted fractions are concentrated throughout the chromatography process.

13. Tetrahydrofuran efficiently removes any residual 1,2-naphthoquinone-4-glycidol that did not successfully graft to the polymer backbone. The final polymer has very limited solubility in tetrahydrofuran, although some product will be removed.

14. The polymer solution is prepared fresh from the dried product for every experiment.

15. The presence of MWCNTs improves the electrochemical reversibility of the NQ species, as well as provides a larger specific surface area, which in turn also improves the physical stability of the bioelectrode architecture.

16. Once the EGDGE is added, it is important to coat the electrodes without any significant delay to prevent undesired hydrolysis of the epoxide groups of the crosslinker.

17. While the FAD-dependent glucose dehydrogenase bioelectrodes are typically dried overnight before use, recent studies have found that bioelectrodes prepared with less-stable enzymes can be adequately dried in 2 h [6].

18. During the initial equilibration step, the current magnitudes associated with the quasi-reversible electrochemistry of the NQ species begins to increase, followed by a slow decrease (until stabilization). This is attributed to the swelling of the NQ-LPEI upon hydration, since the intermolecular distances of the individual NQ species on the polymer backbone will fluctuate upon hydration leading to continuously changing rates of self-exchange (as a function of distance), and thus current densities.

References

1. Meredith MT, Minteer SD (2012) Biofuel cells: enhanced enzymatic bioelectrocatalysis. Annu Rev Anal Chem 5(1):157–179

2. Leech D, Kavanagh P, Schuhmann W (2012) Enzymatic fuel cells: recent progress. Electrochim Acta 84:223–234. doi:10.1016/j.electacta.2012.02.087

3. Barton SC, Gallaway J, Atanassov P (2004) Enzymatic biofuel cells for implantable and microscale devices. Chem Rev 104(10):4867–4886. doi:10.1021/cr020719k

4. Soukharev V, Mano N, Heller A (2004) A four-electron O_2-electroreduction biocatalyst superior to platinum and a biofuel cell operating at 0.88 V. J Am Chem Soc 126(27):8368–8369. doi:10.1021/ja0475510

5. Milton RD, Hickey DP, Abdellaoui S, Lim K, Wu F, Tan B, Minteer SD (2015) Rational design of quinones for high power density biofuel cells. Chem Sci 6(8):4867–4875. doi:10.1039/C5SC01538C

6. Abdellaoui S, Milton RD, Quah T, Minteer SD (2016) NAD-dependent dehydrogenase bioelectrocatalysis: the ability of a naphthoquinone redox polymer to regenerate NAD. Chem Commun 52:1147–1150. doi:10.1039/C5CC09161F

Chapter 16

Layer-by-Layer Assembly of Glucose Oxidase on Carbon Nanotube Modified Electrodes

Alice H. Suroviec

Abstract

The use of enzymatically modified electrodes for the detection of glucose or other non-electrochemically active analytes is becoming increasingly common. Direct heterogeneous electron transfer to glucose oxidase has been shown to be kinetically difficult, which is why electron transfer mediators or indirect detection is usually used for monitoring glucose with electrochemical sensors. It has been found, however, that electrodes modified with single or multi-walled carbon nanotubes (CNTs) demonstrate fast heterogeneous electron transfer kinetics as compared to that found for traditional electrodes. Incorporating CNTs into the assembly of electrochemical glucose sensors, therefore, affords the possibility of facile electron transfer to glucose oxidase, and a more direct determination of glucose. This chapter describes the methods used to use CNTs in a layer-by-layer structure along with glucose oxidase to produce an enzymatically modified electrode with high turnover rates, increased stability and shelf-life.

Key words Enzyme immobilization, Glucose oxidase, Carbon nanotubes

1 Introduction

The use of enzymatically modified electrodes for the detection of glucose or other non-electrochemically active analytes is becoming increasingly common due to the low-cost, real-time operation in addition to a high specificity compared to other detection methods such as chromatography [1–7]. Although the oxidation of glucose oxidase is thermodynamically spontaneous, the distance that the electron has to travel to the electrode surface is far enough that the oxidation of FAD in glucose oxidase is kinetically unfavorable. This makes the heterogeneous oxidation of the FAD in the glucose oxidase an overall unfavorable process by itself. When the electrode is modified with the enzyme as well as a layer-by-layer support system, these modifications do not destroy the enzymatic activity or the enzyme in addition to providing high mechanical stability.

Shelley D. Minteer (ed.), *Enzyme Stabilization and Immobilization: Methods and Protocols*, Methods in Molecular Biology, vol. 1504, DOI 10.1007/978-1-4939-6499-4_16, © Springer Science+Business Media New York 2017

1.1 Electron Transport

Glucose oxidase has become a benchmark enzyme for biosensors. Glucose oxidase has many favorable attributes such as high turnover rate, excellent selectivity, good thermal and pH stability as well as low cost [6]. The active site in glucose oxidase contains the FAD coenzyme molecule which is tightly bound but not covalently attached to the enzyme and fully reversible. Since FAD is buried inside the center of the oxidase, it exhibits inefficient electron transfer to an electrode surface. There are two main methods to overcome this barrier, mediated electron transfer (MET) and direct electron transfer (DET), *see* Fig. 1 [8–11]. In a mediated electron transfer system, an electrochemically reversible redox molecule is used to transport the electron from the active site of the enzyme to the surface of the electrode. MET systems typically use molecules that have fast electron transfer kinetics and an oxidation potential positive of the thermodynamic oxidation potential of the FAD center of the glucose oxidase. Mediators used for these reactions typically use organometallic hydrogel polymers. In these polymers, a redox-active metal center such as osmium, ruthenium, or ferrocene is covalently attached to a polymer backbone like poly(vinylpyridine) (PVP) or poly(ethyleneimine) (PEI) [12, 13]. MET, while making the current densities larger, can affect the stability of the system. To simplify the system but still have large current densities it is possible to produce modified electrodes that directly transfer the electrons from the enzyme to the electrode. Direct electron transport (DET) occurs when an immobilized enzyme active site is both close to the surface and in the correct orientation on the electrode for electron transfer to take place, which overcomes the kinetic barrier to allow heterogeneous oxidation to take place. Carbon nanotubes have been postulated to be able to facilitate electron transfer from glucose oxidase to the surface of the electrode [4].

1.2 Carbon Nanotubes

CNTs are found to have high electrocatalytic effect and fast electron transfer rates. This is due to the chemical and physical properties of these materials. The structure of carbon nanotubes is unique, having graphite sheets along the cylinder with properties similar to the basal plane of highly ordered pyrolytic graphite

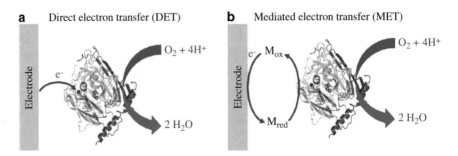

Fig. 1 Electron transfer at an enzymatic electrode using either: (**a**) direct electron transfer or (**b**) mediated electron transfer (Reproduced with permission from Mano et al. [7])

(HOPG). The open ends of CNTs are similar to the edge plane of HOPG. McCreery et al. studied extensively the electrochemical behavior of HOPG electrodes, and have found that that the edge plane has much better electron transfer properties than the basal plan [14]. This means given the much larger proportion of edge to plane with CNTs, it is not surprising that CNTs demonstrate favorable electron transfer kinetics.

Glucose oxidase can be immobilized with CNTs in a variety of ways. Several authors have adsorbed glucose oxidase to carbon nanotubes cast onto carbon electrodes. Guiseppi-Elie et al. reported on electron transfer from CNT modified electrodes to glucose oxidase [15]. They found that glucose oxidase adsorbed to single walled carbon nanotubes cast onto glassy carbon electrodes had current responses characteristic of the redox chemistry of $FAD/FADH_2$. When the glucose oxidase sensor is exposed to glucose in an oxygen saturated solution, a current response is shown that is consistent with the enzymatic conversion of glucose to gluconolactone. The authors speculate that the electron transfer to glucose oxidase is a result of the fibrils of the cast CNT film coming within tunneling distance of the FAD portion of the glucose oxidase enzyme. Like Gooding, Guiseppi-Elie et al. proposed that the single-walled CNTs penetrate the native structure of the enzyme to facilitate electron transfer. Unlike Gooding, Guiseppi-Elie et al. propose that this happens without having to denature and then reconstitute the enzyme structure around the carbon nanotube. However, it is important to note that oxygen is likely producing peroxide in this system.

Another method is to use CNT to immobilize the glucose oxidase to the CNTs using a layer-by-layer assembly [1]. This method allows the close contact between the enzyme and surface of the electrode, and effectively adsorbs the enzyme to the surface of the electrode as well as maintaining a constant level of enzyme on the surface of the electrode [16, 17]. Other approaches like covalent binding or entrapment have been used, but their main drawback is that the distribution of enzyme is not uniform and can change the active site of the enzyme [18]. The layer-by-layer assembly of glucose oxidase electrodes provides a mild environment for the enzyme to be in with increased stability of the film. Layer-by-layer can use a variety of polyelectrolytes to modify the surface, this work uses polydiallyldimethylammonium chloride, but poly(ethyleneimine) or chitosan have also been used.

2   Materials

2.1   Shortening of Nanotubes

1. Single-walled carbon nanotubes.

2. Sonicator.

3. 5 mL centrifuge tubes.

4. Vacuum filtration with 0.45 μm filter paper.

2.2 Preparation of Nanotube Modified Electrodes

1. Glassy carbon electrodes (2 mm diameter).
2. Micropolish.
3. Sonicator.
4. Potentiostat able to perform cyclic voltammetry.
5. PDDA (polydiallyldimethylammonium chloride).
6. Suspension of shortened MWNT in DMF (as described in Subheading 3.1).

2.3 Glucose Stock Solution

1. Phosphate buffer: 0.05 M and pH 7.4.
2. Glucose stock solution: 1.0 M D-glucose solution, prepared and allowed to mutarotate for 24 h (*see* Subheading 3.2).

2.4 Modification of Film Electrodes with Enzyme and Nanotube Modified Electrodes

1. Glucose oxidase solution: Type X-S, 10KU (*Aspergillus niger*, E.C. 1.1.3.4) glucose oxidase at 0.5 mg/mL.
2. HEPES buffer: 0.1 M and pH 7.4.
3. Nanotube modified electrodes (as described in Subheading 3.2).
4. 1 mL centrifuge tubes.

2.5 Determination of Immobilized Enzyme Activity

1. PDDA/GOx/PDDA/CNT film electrodes (*see* Subheading 3.4).
2. Phosphate buffer: 0.1 M and pH 7.4.
3. 2,6-dichlorophenolidophenol (DCPIP): 0.7 mM prepared in water.
4. UV–Vis spectrophotometer.
5. Quartz Cuvettes.

2.6 Electrochemical Testing of Modified Electrodes

1. Phosphate buffer: 0.1 M and pH 7.4.
2. H_2O_2 solution: 100 mM and prepared fresh daily.
3. 1.0 M D-glucose solution, prepared and allowed to mutarotate for 24 h (*see* Subheading 3.2).
4. Potentiostat able to perform cyclic voltammetry and amperometry.
5. Electrochemical cell.
6. Electrochemical flow cell.
7. Reference electrode suitable for aqueous experiments (typically Ag/AgCl).
8. Counter electrode (typically platinum wire).
9. PDDA/GOx/PDDA/CNT film electrodes (*see* Subheading 3.4).

3   Methods

3.1   Shortening of Carbon Nanotubes

Carbon nanotubes have a tendency to aggregate in solution due to their very strong van der Waals interactions [19, 20]. The entanglement due to this aggregation prevents the CNTs from solubilizing in solution easily and do not have any available ends to electrostatically attract enzyme. This method cuts the CNTs and provides free ends that will oxidize and have an overall net negative charge.

1. In a small beaker, add 2 mg of SWNTs and 10 mL of a mixture of 1:3 concentrated sulfuric acid and concentrated nitric acid.
2. Sonicate for 4 h.
3. Cool to room temperature.
4. Dilute with DI water to 500 mL.
5. Filter the resultant solution with a 0.45 μm filter paper.
6. Wash with water until neutral pH is achieved and collect solid.
7. Redisperse in DMF to a concentration of 5 mg/mL.

3.2   Preparation of Nanotube Modified Electrodes

The carbon nanotubes when placed on the surface of the electrode will form a well-ordered layer on the surface. This surface will be negatively charged after rinsing with water and letting dry at ambient conditions.

1. Polish glassy carbon electrodes using alumina slurry.
2. Sonicate electrode in DI water for 10 min to remove all alumina from surface.
3. Place 20 μL of CNT solution onto the cleaned glassy carbon electrode.
4. Allow electrodes to dry at room temperature.
5. The electrodes were then rinsed with DI water and allowed to dry again.

3.3   Glucose Stock Solution

1. Prepare a 1.0 M glucose stock solution by measuring out 180.0 g of glucose into a 1-L volumetric flask.
2. Use 0.1 M phosphate buffer, pH 7.4 to dilute to the line.
3. Let stand to mutarotate for 24 h (see **Note 1**).

3.4   Modification of CNT Film Electrodes with Enzyme

The modification of the glassy carbon electrodes uses a layer-by-layer assembly process. The CNTs are added to the surface and allowed to self-assembly at room temperature. After rinsing with water, the now negatively charged surface has a layer of positively charged PDDA added to the surface. Now that the surface is positively charged, a layer of glucose oxidase (which is negatively charged at pH 7.4) will electrostatically assemble to the surface of the electrode. This is finished with one more layer of PDDA to keep the glucose oxidase on the surface. This procedure specifically uses glucose oxidase, but could also be used with other FAD-dependent enzymes.

1. Prepare a 100 mL 0.5 M NaCl aqueous solution.

2. Dissolve PDDA into the 0.5 M NaCl solution to a final concentration of 1 mg/mL PDDA.

3. In a small 1 mL centrifuge tube, add 500 μL of the 1 mg/mL PDDA solution. Place the CNT modified electrode into the centrifuge tube and let rest at room temperature for 20 min.

4. Rinse the electrode with DI water and let dry.

5. In a small 1 mL centrifuge tube, add a 0.5 mg/mL glucose oxidase solution in 0.1 M HEPES buffer (pH 7.4).

6. Put the PDDA/CNT modified electrode into the centrifuge tube and let rest at room temperature for 20 min.

7. In a small 1 mL centrifuge tube, add 500 μL of the 1 mg/mL PDDA solution and let rest at room temperature for 20 min.

8. Make the second PDDA layer by placing the modified electrode in a small 1 mL centrifuge tube, add 500 μL of the 1 mg/mL PDDA solution and let rest at room temperature for 20 min.

9. Rinse with water and let dry at room temperature.

3.5 Determination of Immobilized Enzyme Activity

After preparing the immobilized enzyme electrodes, it is important to determine if the enzymes still have activity. This is done by comparing the enzyme activity of the prepared biosensor to that of enzyme in free solution. It is well known that immobilization can decrease the activity of the enzyme by limiting the ability of the substrate to reach the active site. However, it is also possible that the immobilization could increase the activity of the enzyme by hindering enzyme. The procedure, described here [21], measures the activity of the immobilized GOx using the electron acceptor DCPIP. DCPIP is blue in its oxidized form and turns colorless when reduced. This can be used in conjunction with the GOx to determine its activity.

1. In a small vial, mix 8.5 mL of 0.1 M phosphate buffer, pH 7.4, 0.5 mL of DCPIP, and 1 mL of glucose solution.

2. Add 2 mL of the mixture to the quartz cuvette.

3. Place the cuvette in the spectrophotometer and measure the absorbance at 555 nm.

4. Suspend the enzymatic electrode in the cuvette so that the modified surface is submerged. Leave the electrode in the solution for 5 min.

5. Remove the electrode and mix by inverting 3–4 times.

6. Place the cuvette in the spectrophotometer and measure the absorbance at 555 nm.

The activity of the immobilized GOx can be calculated from the following equation:

$$\text{Enzyme activity}\left(\frac{\text{Units}}{\text{mg}}\right) = \frac{\left(A_{\text{initial}} - A_{\text{final}}\right)(2\text{mL})}{\left(8.465\,\text{mM}^{-1}\text{cm}^{-1}\right)(5\,\text{min})(1\,\text{cm})(0.005\,\text{mg})}$$

A_{initial} = initial absorbance at 555 nm

A_{final} = absorbance at 555 nm after reaction with enzyme electrode

2 mL = total volume of assay

8.465 mM^{-1} cm^{-1} = millimolar extinction coefficient of DCPIP

5 min = total time of reaction

1 cm = path length of cuvette

0.005 mg = mass of immobilized GOx

One unit of GOx will oxidize 1 μmol of β-D-glucose to D-gluconolactone per minute at room temperature at pH 7.4.

3.6 Electrochemical Testing of Modified Electrodes

Typically two different types of electrochemical experiments are performed to test these types of modified electrodes. The first is cyclic voltammetry (CV) which shows if there is good electrical connection and the other is amperometry which is able to quantify the enzyme activity on an electrode.

3.6.1 Cyclic Voltammetry

In cyclic voltammetry the working electrode is swept through a series of potentials and the resultant current is measured. To determine if the layers are functioning correctly, first the electrode is tested in the absence of glucose (just the buffer). Then the electrode is tested in the presence of hydrogen peroxide. If the electrode is working properly, the hydrogen peroxide will be oxidized and the oxidation peak should increase with increasing H_2O_2 concentration.

1. In an electrochemical cell, place the enzymatic electrode, reference electrode, and counter electrode in phosphate buffer. The volume of buffer will vary based on the size of the container but should be enough to cover the electrodes.

2. Program the potentiostat software with the following parameters (the names may vary based on brand of instrument):

   (a) Low potential: −0.8 V vs. Ag/AgCl

   (b) High potential: 0.2 V vs. Ag/AgCl

   (c) Scan rate: 100 mV/s

   (d) Sweep segments or number of cycles: four segments or two cycles

3. After programming the software with the above parameters, run two cycles with the electrode in buffer.

4. Add enough of the H_2O_2 solution to make the final concentration 10 mM H_2O_2.

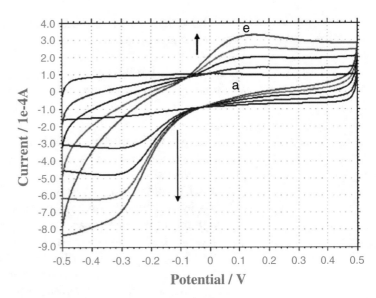

Fig. 2 Cyclic voltammograms of a PDDA/GOx/PDDA/CNT modified electrode in 0.05 M pH 7.4 phosphate buffer containing 0, 1, 2, 3, and 4 mM (from a to e) H_2O_2 at a scan rate of 100 mV/s vs. Ag/AgCl [1]

As can be seen in Fig. 2, a pair of well-defined redox peaks is observed. For a surface-confined electrode reaction a linear plot of scan rate vs. peak potential currents shows that the film is surface bound.

3.6.2 Amperometry

In an amperometry experiment, a fixed potential is applied to the working electrode and the current is measured over time. The potential is chosen based on the cyclic voltammetry data and a potential is chosen that will reduce the effects of electrochemical interferences. The potential is applied and the current initially jumps to a large value before slowly decaying. It should eventually stabilize at a relatively constant value, i.e., steady state current. At this point, substrate can be injected into the solution. After each injection the current should be allowed to stabilize again. For this type of experiment, a flow cell is ideal as it keeps the solution in motion and decreases the noise of the measurement.

1. In a flow cell, place the enzymatic electrode, reference electrode, and counter electrode. The phosphate buffer needs to be continuously flowing without air bubbles.

2. Program the potentiostat software and pump with the following parameters (the names may vary based on brand of instrument):

   (a) Potential: –0.1 V vs. Ag/AgCl.

   (b) Sample interval: how often the instrument takes a data point. Typically one point every second works well.

(c) Flow rate: 0.1 mL/min.

(d) Run time: Sometimes these experiments take a long time so setting the run time to be long is better than too short. The experiment can always be stopped early but for most instruments the parameters cannot be changed after it is already running.

3. After programming the software with the appropriate parameters, begin the experiment.

4. Depending on the electrode, the amount of time it takes for the current to stabilize can vary greatly. Allow a long time for the electrode current to come to a steady state value. Make sure that the phosphate buffer being fed into the flow cell is more than enough for the time of the experiment.

5. When the current has stabilized, begin injections of the glucose solution. Inject the glucose slowly to prevent air bubbles from entering the solution. After each injection allow the current to stabilize before continuing with the next. Again, this time may vary.

With increasing concentration of glucose being injected into the system, the current will increase. As can be seen in Fig. 3 there is increasing current with increasing concentration of glucose. It produces a linear calibration curve as shown in Fig. 4. From this data a Lineweaver–Burk plot can be produced to obtain i_{max} and K_m. These are important parameters to be able to determine the enzyme

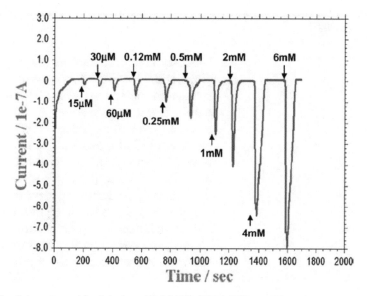

Fig. 3 Amperometric data for a PDDA/GOx/PDDA/CNT modified electrode modified in 0.1 M pH 7.4 phosphate buffer at −0.10 V vs. Ag/AgCl. Various concentrations of glucose injections are noted on the graph [1]

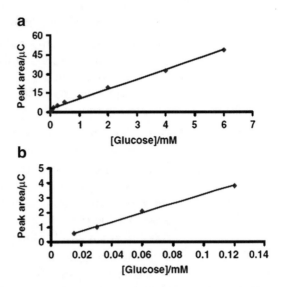

Fig. 4 Calibration curve based on amperometric data for a PDDA/GOx/PDDA/CNT modified electrode over the range of glucose injection concentrations [1]

activity level as well as to compare the immobilization strategy across methods. The K_m gives an estimate of the modified electrode's affinity for glucose and is not an intrinsic property of the enzyme, but of the system as a whole. The smaller the value of the calculated K_m the better the enzyme is functioning in the modified electrode.

4   Notes

1. Glucose standard solutions are required to sit for 24 h after preparation to allow for mutarotation. Mutarotation is the change in optical rotation between two anomers. This is required of a glucose solution since cyclic carbohydrates have an α and β form. D-glucose is typically in the β form. When a glucose solution is made a small percentage of the β form glucose will convert to the α form until an equilibrium ratio between the α and β form is reached. However, glucose oxidase can only oxidize the β form of glucose, so it is critical that the glucose solution is at equilibrium before it is used. If the glucose solution is used for experiments prior to reaching equilibrium, the concentration of the substrate (β-D-glucose) will continue to change over the time span of the experiment.

References

1. Liu G, Lin Y (2006) Amperometric glucose biosensor based on self-assembling glucose oxidase on carbon nanotubes. Electrochem Commun 8(2):251–256. doi:10.1016/j.elecom.2005.11.015

2. Zhu Z, Garcia-Gancedo L, Flewitt AJ, Xie H, Moussy F, Milne WI (2012) A critical review of glucose biosensors based on carbon nanomaterials: carbon nanotubes and graphene. Sensors (Basel) 12(5):5996–6022. doi:10.3390/s120505996

3. Walcarius A, Minteer SD, Wang J, Lin Y, Merkoçi A (2013) Nanomaterials for bio-functionalized electrodes: recent trends. J Mater Chem B 1(38):4878. doi:10.1039/c3tb20881h

4. Harper A, Anderson MR (2010) Electrochemical glucose sensors—developments using electrostatic assembly and carbon nanotubes for biosensor construction. Sensors (Basel) 10(9):8248–8274. doi:10.3390/s100908248

5. Turner AP (2013) Biosensors: sense and sensibility. Chem Soc Rev 42(8):3184–3196. doi:10.1039/c3cs35528d

6. David M, Barsan MM, Florescu M, Brett CMA (2015) Acidic and basic functionalized carbon nanomaterials as electrical bridges in enzyme loaded chitosan/poly(styree sulfonate) self-assembled layer-by-layer glucose biosensors. Electroanalysis 27:2139–2149

7. Mano N, Edembe L (2013) Bilirubin oxidases in bioelectrochemistry: features and recent findings. Biosens Bioelectron 50:478–485. doi:10.1016/j.bios.2013.07.014

8. Calvo EJ, Etcheniqe R, Danilowicz C, Diaz L (1996) Electrical communication between electrodes and enzymes mediated by redox electrodes. Anal Chem 68(23):4186–4193

9. Godet C, Boujtuta M, El Murr N (1999) Direct electron transfer involving a large protein: glucose oxidase. N J Chem 23:795–797

10. Zhang W, Li G (2004) Third-generation biosensors based on the direct electron transfer of proteins. Anal Sci 20:603–609

11. Calabrese Barton S, Callaway J, Atanassov PB (2004) Enzymatic biofuel cells for implantable and microscale devices. Chem Rev 104:4867–4886

12. Fu Y, Zou C, Xie Q, Xu X, Chen C, Deng W, Yao S (2009) Highly sensitive glucose biosensor based on one-pot biochemical pre-oxidation and electropolymerization of 2,5-di mercapto-1,3,4-thiadiazole in glucose oxidase-containing aqueous suspension. J Phys Chem B 113(5):1332–1340

13. Mugweru A, Shen Z (2010) Electrochemistry of protein and redox polymers trapped in polyethylene glycol diacrylate gel. J Undergraduate Res 9(1):1

14. Chen P, McCreery RL (1996) Control of electron transfer kinetics at glassy carbon electrodes by specific surface modification. Anal Chem 68:3958–3965

15. Guiseppi-Elie A, Lei CH, Baughman RH (2002) Direct electron transfer of glucose oxidase on carbon nanotubes. Nanotechnology 13(5):559–564

16. Wu B, Hou S, Miao Z, Zhang C, Ji Y (2015) Layer-by-layer self-assembling gold nanorods and glucose oxidase onto carbon nanotubes functionalized sol-gel matrix for an amperometric glucose biosensor. Nanomaterials 5(3):1544–1555. doi:10.3390/nano5031544

17. Wooten M, Karra S, Zhang M, Gorski W (2014) On the direct electron transfer, sensing, and enzyme activity in the glucose oxidase/carbon nanotubes system. Anal Chem 86(1):752–757. doi:10.1021/ac403250w

18. Ramasamy RP, Luckarift HR, Ivnitski DM, Atanassov PB, Johnson GR (2010) High electrocatalytic activity of tethered multicopper oxidase-carbon nanotube conjugates. Chem Commun 46(33):6045–6047. doi:10.1039/C0CC00911C

19. Liu J, Chou A, Rahmat W, Paddon-Row MN, Gooding JJ (2005) Achieving direct electrical connection to glucose oxidase using aligned single walled carbon nanotube arrays. Electroanalysis 17(1):38–46. doi:10.1002/elan.200403116

20. Wang J (2005) Carbon-nanotube based electrochemical biosensors: a review. Electroanalysis 17(1):7–14. doi:10.1002/elan.200403113

21. Milton RD, Giroud F, Thumser AE, Minteer SD, Slade RC (2013) Hydrogen peroxide produced by glucose oxidase affects the performance of laccase cathodes in glucose/oxygen fuel cells: FAD-dependent glucose dehydrogenase as a replacement. Phys Chem Chem Phys 15(44):19371–19379

Chapter 17

Kinetic Measurements for Enzyme Immobilization

Michael J. Cooney

Abstract

Enzyme kinetics is the study of the chemical reactions that are catalyzed by enzymes, with a focus on their reaction rates. The study of an enzyme's kinetics considers the various stages of activity, reveals the catalytic mechanism of this enzyme, correlates its value to assay conditions, and describes how a drug or a poison might inhibit the enzyme. Victor Henri initially reported that enzyme reactions were initiated by a bond between the enzyme and the substrate. By 1910, Michaelis and Menten were advancing their work by studying the kinetics of an enzyme saccharase which catalyzes the hydrolysis of sucrose into glucose and fructose. They published their analysis and ever since the Michaelis–Menten equation has been used as the standard to describe the kinetics of many enzymes. Unfortunately, soluble enzymes must generally be immobilized to be reused for long times in industrial reactors. In addition, other critical enzyme properties have to be improved like stability, activity, inhibition by reaction products, and selectivity towards non-natural substrates. Immobilization is by far the chosen process to achieve these goals.

Although the Michaelis–Menten approach has been regularly adapted to the analysis of immobilized enzyme activity, its applicability to the immobilized state is limited by the barriers the immobilization matrix places upon the measurement of compounds that are used to model enzyme kinetics. That being said, the estimated value of the Michaelis–Menten coefficients (e.g., V_{max}, K_M) can be used to evaluate effects of immobilization on enzyme activity in the immobilized state when applied in a controlled manner. In this review enzyme activity and kinetics are discussed in the context of the immobilized state, and a few novel protocols are presented that address some of the unique constraints imposed by the immobilization barrier.

Key words Enzyme activity, Enzyme kinetics, Michaelis–Menten, Immobilization

---

1 Introduction

1.1 Enzyme Activity

Enzyme immobilization is used to improve enzyme stability, activity, inhibition by reaction products, and selectivity towards non-natural substrates [1–4]. Enzyme kinetics is the science of using enzyme activity data to model or predict enzyme behavior. Consequently, before enzyme kinetics can be considered, its activity must be measured. Although immobilization can introduce conditions that either alter the enzyme's activity (e.g., through modification of the enzyme structure through physical or chemical forces), or introduce barriers that alter its "apparent" activity

Shelley D. Minteer (ed.), *Enzyme Stabilization and Immobilization: Methods and Protocols*, Methods in Molecular Biology, vol. 1504, DOI 10.1007/978-1-4939-6499-4_17, © Springer Science+Business Media New York 2017

(e.g., through the introduction of mass transfer limitations that limit substrate diffusion from the bulk solution to the enzyme active site), the experimental measurement of activity is the same whether the enzyme is immobilized or freely suspended in solution. With that in mind, a brief summary of the general definition of enzyme activity and how its measured is given although an excellent account can be found elsewhere [5].

1.1.1   Chemical

The international unit for enzyme activity is the International Unit (IU) which is defined as the enzyme activity that converts 1 μmol of substrate per minute. Enzyme activity (as volume units) can be calculated from experimental data a defined in Eq. 1 [5].

$$\text{IU / mL \_ sample} = \frac{\Delta A / s \times \text{test\_volume}(\text{mL}) \times df \times 10,000}{\varepsilon_{nm}\left(1 \times \text{mol}^{-1}\text{cm}^{-1}\right) \times \text{enzyme\_volume}(\text{mL})} \quad (1)$$

Where *df* is the dilution factor and 10,000 is a unit conversion. When multiplied by the total enzyme volume Eq. 1 yields the total enzyme activity. When divided by the total protein in the solution (as measured by protein assay), the specific enzyme activity is achieved (Units/mg or Units/g).

The key measurement, $\Delta A/s$, effectively represents the amount of product produced per unit time. For these considerations it is assumed that the time course of the enzyme reaction can be followed continuously, e.g., via optical test or other spectroscopic techniques wherein the initial rate can be calculated using the tangent method to estimate the slope. In the case of oxidative/reductive enzymes, the flow of electrons can be monitored continuously using current measurement or the monitoring (e.g., via absorbance) of a coupled mediator compound that is reduced during the reaction [6]. Given that the use of initial rates is a requirement of the Michaelis–Menten approach to model enzyme kinetics (discussed below), it should be noted that Eq. 1 does not specify a substrate concentration. It simply defines the activity measurement for any given set of experimental conditions including the concentration of the rate limiting substrate.

The product formed can also be measured at a defined end point where, presumably, the reaction is stopped by action of adding an inhibiting agent [7–12]. The assay is stopped after a defined period of time and the final concentration of product measured. The production rate is based upon the difference between the single value and the blank [5].

There are some subtle differences between the initial rate and end point methods that can become particularly relevant when applied to immobilized systems. For example, if the substrate concentration is not maintained constant, or if there is significant mass transfer resistance, $\Delta A/t$ will not increase linearly with time but will more than likely decrease with time. As such the activities that are calculated should be considered an averaged value and must therefore be discussed in the context of the specific assays conditions and relative comparisons must be made from data taken under exact conditions. There also exists the issue that specific events, such as inhibition will not be realized. For example, consider a plot of reaction product versus time (Fig. 1). A typical enzyme reaction might follow the path of the solid line. By contrast, an enzyme that is inhibited (or enzyme reaction that is mass transfer limited) might follow the path of the dashed line. Measuring the amount of product produced at time point A (initial rate conditions) would clearly reveal the inhibition. Measuring the amount of product produced at time point B (end point conditions) would suggest that the enzyme is being inhibited, but not to the same extent as identified under initial rate conditions.

Finally, the activity can be measured as a constant rate [13]. This method utilizes flow cells in which fresh medium is passed over or through the immobilization matrix. In some cases, medium is recycled against fresh medium at a 20 or 40 to 1 ratio [14–16]. This approach, which is confined to immobilized enzyme, yields a single activity measurement for a given substrate concentration in the flow medium. Additional data for enzyme kinetic analysis is obtained by modifying the flow rate, under constant substrate feed concentration, or modifying the feed substrate concentration under conditions of constant feed rate.

1.1.2 Electrochemical

Electrochemical measurement of enzyme activity is a subset limited to enzymes catalyzing oxidation/reduction reactions. Rather than

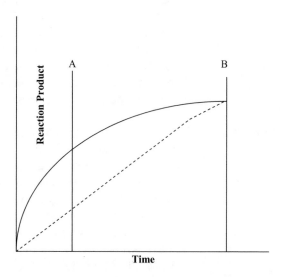

Fig. 1 Plot of measured reaction product versus time for typical enzyme catalyzed reaction in which the reaction product concentration is measured directly (e.g., HPLC) or indirectly (e.g., O.D.)

defining and modeling activity in terms of the rate at which a chemical compound is produced per unit time, the activity is characterized in terms of the rate at which electrons are captured and released from the oxidized substrate. Clearly, a mechanism to count the electrons released is required and this is accomplished either through mediated electron transfer or direct electron transfer. In mediated electron transfer the electrons released in the oxidation of the substrate are transferred from the oxidized substrate to a mediator molecule (such as the $NAD^+/NADH$ system) which then diffuses to the surface of the working electrode where it is oxidized [17]. The released electrons then flow from the mediator (at the electrode surface) through an external circuit where they are counted as current. In the case of direct electron transfer, the electrons released from the oxidized substrate are transferred directly to the electrode surface [18]. Other more complex coupled reactions also occur but they are beyond the scope of this discussion.

In either case the working electrode and enzyme are placed in a well stirred batch type apparatus filled with appropriate buffer/reaction solution along with the appropriate reference and counter electrodes. In the case of immobilized enzyme, the immobilization matrix will be fixed to the electrode surface. The electrode surface will then be set to predefined potential (i.e., in the case of mediated electron transfer the potential will be that required to oxidize the reduced mediator), the substrate added, and the current measured. This is somewhat analogous to the initial rate method.

An alternative method is to apply cyclic voltammograms (CV) over a voltage range that spans both the oxidative and reductive reactions. The peak current is then taken from the oxidative path of the CV. This is somewhat analogous to the endpoint method since the oxidative/reductive reactions bring the oxidation reduction reaction to an equilibrium value for the given scan rate.

Finally, if a flow cell is used the peak current can be measured at steady state and as a function of flow rate. Alternative, a rotating disk electrode can be used and the limiting current can be measured as a function of rotational speed [19].

1.2 Enzyme Kinetics

1.2.1 Chemical

The discussion above considered only a single measurement of activity, as defined by the definition of an IU, and as measured by method of initial rate, end point, or constant rate. It is useful to note that Eq. 1 does not facilitate insight into the kinetics of enzyme behavior, or how to describe, in physiochemical terms, why a change in activity may or may not occur. To that end enzyme kinetics is a method to mathematically describe enzyme activity. The best known example of enzyme kinetic analysis is the method of Michaelis–Menten [20]. Although the best known kinetic model used to predict enzyme activity, it remains limited to situations where simple and straightforward kinetics can be assumed—i.e., there is no intermediate or product inhibition, and there is no allostericity or cooperativity). More, it was developed for irreversible single-substrate reactions despite the reality that most enzyme reactions must be regarded as reversible and using two or more substrates as well as cofactors or coenzymes [5]. Because such considerations, however, quickly lead to complicated relationships, simplifications are often introduced into the experimental protocol (e.g., limitation to initial rates, maintaining all components other than a single rate limiting substrate as constant).

Figure 2 depicts the typical Michaelis–Menten plot of measured reaction velocities as a function of substrate concentration. Each data point in this curve represents an individual experiment (whether it be determined by initial rate, end point, or constant rate) in which the rate of product production is measured and plotted with respect to time. The inset shows an example of an initial rate experiment wherein the initial velocity for a given concentration of the rate limiting substrate S is computed from the initial linear region of the curve (i.e., $\nu_I = dP/dt = -dS/dt$). Additional experiments are conducted at varying concentrations of S and the resultant ν–S plot is then fitted to the Michaelis–Menten model from which V_{max} and K_M are best determined by a second plot of the data in the form proposed by Lineweaver and Burk (i.e., $1/V$

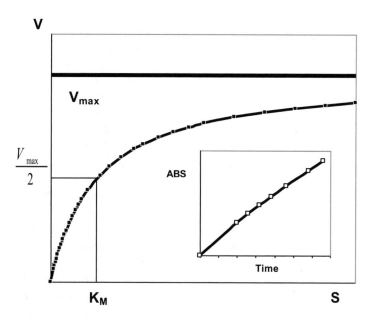

Fig. 2 Michaelis–Menten representation of enzyme activity data. V represents the initial velocity rate of a given enzyme reaction and is calculated from the initial linear portion of the measurement as depicted in the *inset*. S is the limiting substrate concentration for each rate experiment. ABS represents a typical measurement of the reaction product being measured, usually measured by UV/VIS and termed the absorbance (ABS)

is plotted versus $1/S$ and the slope is taken to be K_M/V_{max} and the intercept as $1/V_{max}$) [21].

The initial rate becomes visible (there is actually an initial "first phase" part of this curve that is too rapid to be measured) as the steepest part of the curve (Fig. 1, inset). Because of the inherent assumption of zero order, the slope (or initial rate) must be taken from the linear part of the curve. The main problem with evaluation of initial rates is visualization of the linear portion of the curve (i.e., where dP/dt and dS/dt equal a constant). If the linearity cannot be detected, the Michaelis–Menten equation cannot be used to model the activity data under the assay conditions. It is important to note that this demand on linearity holds for all concentrations of the rate limiting substrate S applied and not just saturating values of S.

A common reason why the linear initial-rate region is difficult to detect is because the assay conditions are not optimum. The reality, however, may be the exact opposite: rather than too little active enzyme, Michaelis–Menten conditions are more likely to hold when less active enzyme is present. Thus, when linear cannot be achieved the initial rate experiments should first be repeated with less immobilized enzyme. And if possible, the sensitivity of detection should be increased. Finally, if decreased enzyme does not produce a linear region, attention should be dedicated to ensuring that pH, ionic strength, and temperature are all well maintained in the enzyme microenvironment. The molarity of the buffer is important and frequently a value of 0.1 M is chosen (which is actually low compared with the concentrated cell solution [5]). The nature of the ions can also destabilize an enzyme and the most appropriate buffer for the enzyme should be selected and this should be determined with assays on the enzyme in free solution prior to its use in assays on the immobilized enzyme. Finally, enhanced mass transfer of all compounds through the application of forced flow (e.g., either through stirring of the solution above the immobilization matrix, using a rotating disk electrode, or passing the fluid through or over the immobilization matrix in a flow cell) should be applied to limit variations in the local pH, ionic strength, and temperature. The strength of the buffer system should also not be overlooked.

1.2.2 Electrochemical

Since a proportional amount of electrons are released per substrate molecule, some analogy with the chemical measurement methodology is appropriate. If the peak current is measured for additional substrate concentrations, the data can be plotted similar to a Michaelis–Menten plot except that rather than plotting the initial rates versus substrate concentration, the peak current is plotted as a function of substrate concentration. Figure 3 shows such a plot for several cases of the enzyme malate dehydrogenase immobilized in different forms of the same polymer.

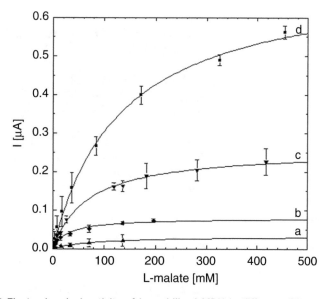

Fig. 3 Electrochemical activity of immobilized MDH in different chitosan polymers. Legend: *Filled squares* (*d*): C4-Chitosan scaffold; *filled* and *inverted triangles* (*c*): C4-Chitosan film; *Filled diamonds* (*b*): Chitosan film; *filled triangles* (*a*): ALA-Chitosan film

For any given curve the current can then be fitted to the Michaelis-Menten "like" constants I_{max} and K'_M [22].

$$I = \frac{I_{max} \cdot S}{S + K_M^{app}} \qquad (2)$$

or

$$I = I_{max} + {}^{c}\!\!\bigg/\!\!\left(K_M^{app} + c\right)$$

Where I_{max} is the maximum obtainable current. Although the Michaelis–Menten theory is not precisely transferable, this process is useful when comparing across various immobilization strategies. For example, Fig. 3 presents several "Michaelis–Menten" curves for the enzyme malate dehydrogenase immobilized in various forms of modified chitosan polymer [23–26]. For each case a different value of I_{max} and K'_M are determined, and this data can be used to infer (at least to some degree of physiochemical accuracy) the effects of immobilization. For example, Table 1 shows model fits to the data presented in Fig. 3. Although exact meanings to the differences in I_{max} and K_M^{app} are not readily obvious, one can infer effects of enzyme activity in terms of the enzyme's ability to work (I_{max}) or its affinity for the substrate (K_M^{app}).

Alternatively the reaction rate can be directly correlated to the measured current through Faraday's constant [27].

Table 1
Amperometric data of I_{max} and resulting K_M^{app} for each chitosan thin film and scaffold tested

	I_{max} (nA)	K_M^{app} (mM)
ALA film	33.4 ± 5.6	81.7 ± 27.4
CHIT film	80.1 ± 4.1	30.1 ± 4.9
C4 film	257.4 ± 7.4	69.2 ± 6.6
C4 scaffold	708.8 ± 21.1	132.7 ± 10.5

$$v = \frac{I}{n \cdot F \cdot A} \tag{3}$$

where v is the reaction rate defined as moles of electrons released in the oxidation reaction per unit area per unit time (mol/s/cm²), I is the measured current defined as electrical charge in coulombs per unit time (C/s), F the Faraday constant is the magnitude of electrical charge in coulombs per mole of electrons (C/mol), A the area of the electrode (cm⁻²), n the number of electrons transferred. If one knows the moles of substrate consumed, or product produce for each mole of electrons released, the reaction rate in terms of moles of production is easily calculated.

In the case of immobilized enzymes, it is particularly useful to explore diffusion effects. To do this, the peak current from CV plots can be plotted versus the square root of the scan rate. Specifically the CV's are repeated for a single substrate concentrations under identical conditions except the scan rate is changed (it is also assumed that the concentration of the substrate does not change during the CV's). If all cyclic voltammetric experiments yield a liner I_p versus $v^{1/2}$ plot, transport-limited electrochemistry can be assumed. More, the flux through the immobilization matrix can be estimated [25].

$$Flux = \frac{I_p}{n \cdot F \cdot A} \tag{4}$$

Where $I_p = 2.69 \times 10^5 \cdot n^{3/2} \cdot AC^* \cdot v^{1/2} \cdot KD^{1/2}$ and I_p is peak current, n the number of electrons transferred, F the Faraday constant, A the area of the electrode, C^* the concentration of redox species, v is the scan rate, K the extraction coefficient, and D is the diffusion coefficient.

Diffusion effects can also be investigated with the rotating disk electrode wherein the peak current is measured for a given concentration at various rotational rates [19]. In other cases, CV's can be run versus a number of fixed rotational speed rates [28].

2   Materials

2.1   Enzyme Characterization Using Flow Through Cells [13, 29]

1. Alcohol dehydrogenase (EC 1.1.1.1, CAS 9031-72-5) stored at −20 °C.

2. β-Nicotinamide adenine dinucleotide (NAD⁺) and β-Nicotinamide adenine dinucleotide, reduced dipotassium salt (NADH) stored at −20 °C. Assay the concentration of spectrophotometrically active NAD⁺. Run the alcohol dehydrogenase catalyzed reaction to completion. Assay the concentration of NADH at 340 nm ($\varepsilon = 6300$ M⁻¹ cm⁻¹). Note: the actual concentration of spectrophotometrically active NAD⁺ is calculated from Beer's law using an extinction coefficient of 6.3 mM⁻¹ cm⁻¹.

3. NAD⁺ solution in phosphate buffer: 3.0 mM NAD⁺ (2.16 mg solid per ml, assuming a mass purity of 92 %) in a solution containing phosphate buffer (0.05 M pH 8.0), ethanol (0.1–0.3 M), ADH (0.1 mg/ml, diluted from stock solution), and BSA (1.0 mg/ml). BSA is added as a protective agent against enzyme degradation and aggregation (*see* **Note 1**). Keep on ice and use within 6 h to avoid degradation as per manufacturer's recommendations.

4. pH 8.0 phosphate buffer: Make phosphate buffer by mixing separate stock solutions of 0.2 M K_2HPO_4 and KH_2PO_4, in proportions necessary to achieve a final pH of 8.0 via calibrated pH probe. The final stock solution is diluted to 0.05 M and filtered through 0.2 μm filters, and deaerated with N_2 gas.

5. Flow-through electrochemical cell with reference (Ag/AgCl, CH Instruments) and counter electrode (platinum wire, Alfa-Aesar)) [6, 29].

6. UV/VIS spectrophotometer with quartz flow-through cell.

7. Peristaltic pump.

8. BSA stock solution: BSA dissolved in stock solutions (100 mg/ml) of phosphate buffer (pH 8.0).

2.2   Amperometric Characterization of Immobilized Enzyme on Glassy Carbon [30]

1. Methylene green solution: 0.4 mM reagent grade methylene green and 0.1 M sodium nitrate in 10 mM sodium borate solution.

2. Medium molecular weight chitosan (MMW): Deacylate native medium molecular weight chitosan by autoclaving at 121 °C for 2 h in 40 wt% NaOH solution. Filter wash the autoclaved residue with DI water and phosphate buffer (pH 8) before

drying in a vacuum oven at 50 °C for 24 h [31]. This will achieve a final deacetylation degree of 95 % (*see* **Note 2**).

3. CHIT stock solution: Prepare a 1 wt% stock solution of deacylated chitosan (CHIT) in 0.25 M acetic acid. Store at room temperature under mild stirring.

4. Prepare hydrophobically modified chitosan in the form butylchitosan (C4-CHIT) and α-linoleic acid-chitosan (ALA-CHIT) from the deacylated chitosan as originally presented elsewhere [24, 26]. For C4-CHIT, specifically, 0.5 g of medium molecular weight chitosan was dissolved in 15 ml of 1 % acetic acid solution under rapid stirring until a viscous gel-like solution was achieved. 15 ml of methanol was then added and the mixture allowed to stir for an additional 15 min at which time 20 ml of butyraldehyde was added, followed immediately by addition of 1.25 g of sodium cyanoborohydride. The gel-like solution was continuously stirred until the suspension cooled to room temperature. The resulting product was separated by vacuum filtration and washed with 150 ml increments of methanol. The hydrophobically modified chitosan was then dried *in vacuum* at 40 °C for 2 h, leaving a flaky white solid absent of any residual smell of aldehyde. A portion of this polymer was then suspended in 0.2 M acetic acid to create a 1 wt% solution and vortexed for 1 h in the presence of 2 and 5 mm diameter yttria stabilized zirconia oxide balls (Norstone, Wyncote, PA) [25]. For ALA-CHIT, 1 g of MMW chitosan was dissolved in 100 ml of 0.2 M acetic acid and 85 ml of methanol under rapid stirring until a viscous solution was achieved. 200 μl of α-linoleic acid was then added to the solution. 10.5 mg of EDC (dissolved in 15 ml methanol) was then added; drop wise, to the chitosan solution. The reaction was then covered and left to stir on a magnetic stirrer plate for 24 h. The resulting reaction mixture was then halted by pouring the solution into 200 ml of a methanol/ammonia solution (7:3 v/v) under constant stirring. The precipitate was the rinsed with 1 L of distilled water followed by 500 ml of methanol and then 200 ml of ethanol over a vacuum filtration system. The ALA-CHIT was then dried *in vacuum* at 40 °C for 1 week, yielding a flaky yellow solid [32].

5. Modified chitosan solution: Prepare 1 wt% polymer stock solutions of each modified polymer in 0.25 M acetic acid and store at room temperature until used.

6. Malate dehydrogenase (porcine heart, lyophilized powder, No. 18670, USB Corporation, Cleveland, OH USA).

7. β-Nicotinamide adenine dinucleotide (NAD⁺) and β-Nicotinamide adenine dinucleotide, reduced dipotassium salt (NADH).

8. L-(−) Malic acid.

9. Glassy carbon (CH Instruments, # CHI104) and Ag/AgCl reference electrodes (CH Instruments, # CHI111).

2.3 Charge Transfer Efficiency of Immobilized Enzyme [6]

1. Hydrogenase enzyme or similar oxidative/reductive enzyme that produces a spectrophotometrically active co-product.

2. Buffer: 0.05 M KH_2PO_4, 0.1 M KCl, pH 7.

3. 25 mm × 4.5 mm strip of platinum foil (Alfa Aesar, MA, USA).

4. Miniature Ag/AgCl type reference electrode (BAS, West Lafayette, IN USA).

5. Graphitic carbon (Superior Graphite, IL, USA).

6. Methyl viologen.

7. 316 stainless steel mesh disks (total height or thickness of 1.5 mm) were cut into 5.6 mm diameter disks geometric surface areas of 0.25 cm^2.

8. pH 7 phosphate buffer: 0.05 M KH_2PO_4, 0.1 M KCl, pH 7.

9. N_2 or H_2 gas.

10. Glove box (if needed for enzymes that must be in O_2-free atmosphere).

11. Potentiostat Gamry 500 potentiostat or equivalent.

3 Methods

3.1 Enzyme Characterization Using Flow Through Cells [6, 29]

1. Place surface holding immobilized enzyme into flow cell (e.g., physical absorption or polymer entrapment to either a carbon felt material or 316 stainless steel wire mesh (Fuel Cell Store, Boulder, CO)).

2. Fill cell to capacity with phosphate buffer solution (0.05 M, pH 8.0).

3. Slowly pump buffer solution (20 ml) through the flow cell (~1.6 ml/min) to purge the cell of non-adsorbed enzyme.

4. Pump feed solution of 3.0 mM NAD+.

5. Spectrophotometrically measure NADH concentration in the outflow.

6. Measure the left hand side (LHS) of Eq. 5 experimentally and plot the right hand side (RHS) in the form of a Lineweaver–Burk plot, assuming a value for the combined mass transfer parameter M_{NAD^+}. Equation 5 has three unknowns: $V_{max}^{NAD^+}$ (μmol/min), $K_M^{NAD^+}$ (mM) and M_{NAD^+} (cm^3/min). Picking a value for M_{NAD^+} allows one to plot the data in Lineweaver–Burk form and facilitates solving for $V_{max}^{NAD^+}$ and $K_M^{NAD^+}$ as per the intercepts (Fig. 4). Their values, however, are arbitrarily dictated by the choice of M_{NAD^+} and therefore potentially

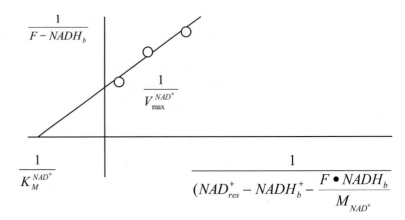

Fig. 4 Graphical depiction of the Lineweaver–Burk solution to Eq. 5

meaningless. If, however, the assumption is made that the value for $K_M^{NAD^+}$ will not change from the K_M measured for the enzyme in free solution, then a value of M_{NAD^+} can be selected to yield a value for $K_M^{NAD^+}$ that is equivalent to the value determined in free solution. The value for V_{max} is then read off the intercept at the y-axis. Making this correction permits derivation of a V_{max} for the electrode that is separated from the effects of mass transfer.

$$F \cdot NADH_b = \frac{V_{max}^{NAD^+} \cdot \left(NAD_{res}^+ - NADH_b - \dfrac{F \cdot NADH_b}{M_{NAD^+}} \right)}{K_M^{NAD^+} + \left(NAD_{res}^+ - NADH_b^+ - \dfrac{F \cdot NADH_b}{M_{NAD^+}} \right)} \quad (5)$$

7. If one further assumes that the immobilization process does not affect the turnover rate of the immobilized enzyme (relative to its activity in free solution), then this value of V_{max}, which represents the total activity of all bound enzyme, can be used to estimate the amount of enzyme that was immobilized. This can be useful when fabricating electrodes using immobilization techniques that entrap a fraction of enzyme from global solutions (such as direct absorption or co-immobilization of gels from free solution) (*see* **Note 3**).

3.2 Amperometric Characterization of Immobilized Enzyme on Glassy Carbon Electrode [30]

1. Prepare poly(MG) coated glassy carbon electrodes as originally described elsewhere [33]. Polish 3 mm glassy carbon electrodes on a Buehler polishing cloth with 0.05 μm alumina. Rinse polished electrode with 18 MΩ water. To form a thin film of poly(MG), electropolymerize the electrode in the presence of methylene green solution using cyclic voltammetry (−0.3 to

1.3 V versus a Ag/AgCl reference electrode for ten scans at a scan rate of 0.05 V/s). The polymerized electrode is then rinsed with 18 MΩ water and allowed to dry in air over night.

2. Coat the poly(MG) electrode with immobilized enzyme. The enzyme/polymer casting solution is made by adding 1 mg of lyophilized malate dehydrogenase to 100 μl polymer solution (1 wt% polymer in 0.25 M acetic acid, containing 1 mg/ml NAD$^+$). The entire mixture is then vortexed for 30 min at +4 °C and 10 μl of the resulting mixture is then pipetted onto the poly(MG) coated electrode [34]. For the preparation of air-dried films, the respective casting solutions are air-dried (3 h) at room temperature under air circulation. For preparation of scaffold electrodes, the electrode plus casting solutions are immediately frozen at –80 °C for 1 h and then freeze-dried in a commercial freeze dryer (3 h). Because of some stability problems with the butyl-modified chitosan scaffolds, those scaffold electrodes should be prepared using a chitosan/butyl-chitosan solution prepared at a ratio of 1:1. All dried anodes are stored at –20 °C and tested within 24 h of preparation. Prior to electrochemical characterization, the dried electrodes should be soaked in 3 M (NH$_4$)$_2$SO$_4$ solution (10 min) and carefully rinsed for 1 min with sodium phosphate buffer (pH 8) (*see* **Note 4**).

3. Execute amperometry using a three-electrode arrangement (Pt-counter electrode, Ag/AgCl reference electrode and glassy carbon/PMG/MDH working electrode) at a constant potential of +300 mV in 10 ml phosphate buffer (pH 7.4). Pipette aliquots of 2 M L-malate solution (pH 7.4 phosphate buffer) into the stirred solution and measure the resulting current measured with time. For each aliquot addition, there is an associated concentration of L-malate in solution. For each concentration (assumed not to change if the solution volume is quite large relative to the surface area of the electrode and thus minimal amounts of the L-malate are consumed) of L-malate, a steady state current will be achieved. This current is then plotted (*y*-axis) against the L-malate concentration (*x*-axis) for a number of L-malate concentrations (Fig. 3). If enough measurements are taken, a Michaelis–Menten type curve will be obtained from which parameters such as I_{max} and K_M' can be determined from simple curve fitting to Eq. 2 (for example). This process can be repeated for various immobilization polymers as presented in Fig. 3 and in this manner their impact on enzyme immobilization discussed in relative terms.

3.3 Charge Transfer of Immobilized Enzyme [6]

1. Obtain suitable hydrogenase enzyme (*see* **Note 5**).

2. Dilute enzyme solution to 2 mg of protein/ml in N$_2$-degassed buffer.

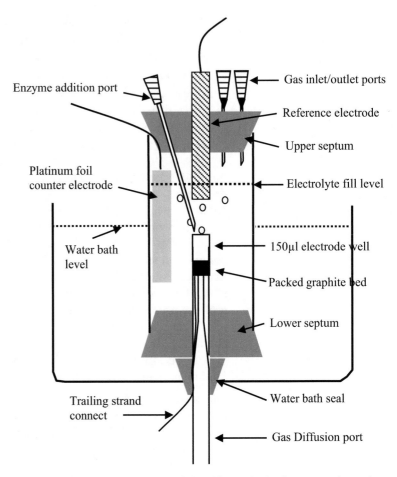

Enzyme addition port

Gas inlet/outlet ports

Reference electrode

Upper septum

Platinum foil counter electrode

Electrolyte fill level

Water bath level

150µl electrode well

Packed graphite bed

Lower septum

Trailing strand connect

Water bath seal

Gas Diffusion port

Fig. 5 Mini-electrochemical cell configured for packed column type electrodes

3. Fabricate three-dimensional stainless steel supports confined within a glass tube (OD 7.2 mm, ID 5.6 mm, Fig. 5). Pack two adjacent 316 stainless steel mesh disks (total height or thickness of 1.5 mm) of 5.6 mm diameter into glass tube. The upper disk should possess a trailing strand for electrical connection. The tapered end of the glass pipette is inverted and sealed with epoxy resin into the lower base, both to seal the electrode end and to provide an upward gas diffusion pathway for hydrogen gas. The glass tube is then placed within a second, larger glass tube that is sealed with a rubber septum to allow submersion into a temperature (75 °C) controlled water bath (mini-gas diffusion cell, Fig. 5). The SS/PGC electrode has a geometric surface area of 0.25 cm². The counter electrode is a platinum foil strip (25 mm×4.5 mm) entering through the top of the cell, and the reference electrode a miniature Ag/AgCl type similarly placed.

4. Entrap the graphite powder (Superior Graphite, IL, USA) onto the stainless steel supports: The graphite powder is first pretreated with concentrated sulfuric acid for 20 min, rinsed copiously in pH 7 phosphate buffer, and then suspended within a distilled water slurry. Approximately 5 ml of this slurry is then pipetted onto the top of the metal support (i.e., stainless steel disks) and pushed through the metal mesh under pressure applied with a 10 ml syringe. The electrode gas entry port (underneath the electrode) is left open during this period to allow slurry to completely traverse the nickel foam or stainless steel mesh and thus coat the entire surface. After drying at 130 °C for 1 h, excess carbon is removed, leaving a 150 μl well space between the surface of the electrode and the upper rim of the glass tube (Fig. 5).

5. Immobilization enzyme onto the packed graphite carbon electrode: Purge the cell atmosphere continuously with hydrogen (10 ml/min) through the upper stopper via a 20-gauge needle while the cell is kept partially filled (i.e., beneath the 150 μl electrode well) with electrolyte to facilitate heat transfer from the water-bath. Syringed 50 μl of enzyme solution (2 mg/ml) into the 150 μl electrode well space using the enzyme addition port (a 16-gauge needle sleeve) orientated to deliver solution directly into the well. The enzyme solution should be left for 2 h at 75 °C, after which time it should be removed using an HPLC syringe inserted through the enzyme addition port. At this time, the mini-diffusion cell is filled to the 3 ml level with electrolyte, and gas purging is switched to the electrode gas addition port in order to provide diffusive gas flow directly through the electrode substrate.

6. Potentiostatic DC polarization is used to characterize the performance of the electrodes (with or without enzyme) in a three-electrode half-cell configuration (*see* **Note 6**). The enzyme and its electrode support as a working electrode (WE) must be held at a constant polarization potential against the reference electrode (RE), and current flowing from the WE to platinum counter electrode (CE) is then monitored. Measured current is taken after the potential step on the electrode is imposed and stabilized for 1 h.

7. The entire cell is then filled with degassed electrolyte (3 ml), in order to allow contact with the platinum counter electrode.

8. Spectrophotometric detection of hydrogenase (both bound and unbound) is determined by measuring the appearance of reduced methyl viologen at 580 nm [35]. Specifically, samples are incubated in hydrogen saturated EFC buffer and the color change (measured as a change in absorbance) is used to monitor the rate of hydrogenase-catalyzed hydrogen oxidation and reduction of methyl viologen. For elevated temperature opera-

tion (70–100 °C), sealed cuvettes should be incubated in a temperature-controlled water bath prior to their transfer to the spectrophotometer. Cuvettes for bound hydrogenase determination should comprise a sealed screw cap to allow electrodes to be immersed in the solution and a micro-stirrer to circulate electrolyte across the electrode.

9. The calculation of bound and free enzyme are similar. In both cases a value for absorbance increase (determined from linear regression of absorbance increase with time) is converted to units of enzyme activity using Beer's law (Eq. 6) and the unit definition of activity for hydrogenase enzyme. Using Beer's law the molar rate of methyl viologen reduced (mM/min) per unit volume (mol/l/min) was calculated as:

$$\frac{\Delta C}{\Delta t} = \frac{\Delta A}{\Delta t} \cdot \frac{1}{\varepsilon \cdot l} \tag{6}$$

where $\Delta C/\Delta t$ is the rate at which methyl viologen is reduced (mol/l/min), $\Delta A/\Delta t$ is the slope of the measured absorbance change with time (min^{-1}), ε is the extinction coefficient of methyl viologen (9700 M^{-1} cm^{-1}), and l is the path length (1 cm). To convert to units of enzyme activity per unit volume, Eq. 6 is multiplied by the definition of activity for hydrogenase enzyme (1 unit of enzyme activity equals 2 μmol methyl viologen reduced per minute)

$$U = \frac{\Delta C}{\Delta t} \cdot \frac{1}{2} \frac{\text{unit} \cdot \text{enzyme} \cdot \text{activity}}{\mu M_{MV} \, / \, \text{min}} \tag{7}$$

where U is the measured activity per unit volume (U/l) of enzyme solution in the cuvette. To obtain a value of the activity per unit electrode surface area (U/cm) Eq. 7 is multiplied by the volume of solution in the cuvette and then divided by the geometric surface area of the electrode.

10. To calculate charge transfer efficiency of bound enzyme, a theoretical maximum in current density is first calculated from spectrophotometrically determined enzyme loading (assuming perfect charge transfer efficiency), and then compared to the actual measured current density obtained from the DC polarization experiment. The maximum theoretical current density is calculated from Faraday's law:

$$I = n \cdot F \cdot \frac{\mathrm{d}H \, / \, \mathrm{d}t}{A} \tag{8}$$

where I is the current density (mA/cm), $\mathrm{d}H/\mathrm{d}t$ is the hydrogen oxidation rate (mmol/s), A is the electrode geometric surface area (cm^2), F is the Faraday constant (96,485 C/mol), and n is the moles electrons released per mole of hydrogen oxidized (Eq. 2). The hydrogen oxidation rate per unit electrode surface area $(dH/dt)/A$ can be determined by multiplying the activity per unit electrode surface area (U/cm) by the definition of hydrogenase enzyme activity (1 unit of enzyme activity is equal to 1 μmol of hydrogen oxidized per minute). The charge transfer efficiency of bound enzyme is then the ratio of the actual measured current density to the theoretical maximum, expressed as a percentage.

4 Notes

1. There is a time dependency for the efficacy of this solution. NAD$^+$ and alcohol dehydrogenase are only stable in solution for 6–8 h while ethanol at these concentration will evaporate from open containers.

2. This step is done, because commercially available chitosan has batch dependencies that result in degrees of deacetylation from 70 to 95 %. Consequently this step is executed to ensure consistency in the degree of chitosan deacetylation from batch to batch and supplier to supplier.

3. This analysis assumes no leaching of immobilized enzyme on the time scale of this experiment.

4. The purpose and importance of this soaking step is to re-equilibrate the enzymes to a salt solution that is used by the manufacturer to stabilize their activity during transport and long term storage.

5. For details on hydrogenase enzyme purification, see as described elsewhere [6]. Otherwise, use suitable oxidative reductive enzyme.

6. Potentiostatic DC polarization and amperometry are identical. While the potentiostat applies a constant potential and measures current, analytical chemists describe the technique as amperometry while chemical engineers term it potentiostatic DC polarization.

References

1. D'Souza SF (2001) Immobilization and stabilization of biomaterials for biosensor applications. Appl Biochem Biotechnol 96(1–3):225–238

2. Gianfreda L, Scarfi MR (1991) Enzyme stabilization: state of the art. Mol Cell Biochem 100(2):97–128

3. Mateo C, Palomo JM, Fernandez-Lorente G, Guisan JM, Fernandez-Lafuente R (2007) Improvement of enzyme activity, stability and selectivity via immobilization techniques. Enz Microb Technol 40(6):1451–1463

4. Monsan P, Combes D (1988) Enzyme stabilization by immobilization. Methods Enzymol 137:584–598

5. Bisswanger H (2004) Practical enzymology. Wiley-VCH Verlag GmbH & Co. KGaA, Weinheim

6. Johnston W, Cooney MJ, Liaw BY, Sapra R, Adams MWW (2005) Design and characterization of redox enzyme electrodes: new perspectives on established techniques with application to an extremophilic hydrogenase. Enz Microb Technol 36:540–549

7. Breton-Maintier C, Mayer R, Richard-Molard D (1992) Labeled proteinase inhibitors: versatile tools for the characterization of serine proteinases in solid-phase assays. Enz Microb Technol 14:819–824

8. Olsson L, Mandenius CF (1991) Immobilization of pyranose oxidase (Phanerochaete chrysosporium): characterization of the enzymic properties. Enz Microb Technol 13:755–759

9. Rani AS, Das MLM, Satyanarayana S (2000) Preparation and characterization of amyloglucosidase adsorbed on activated charcoal. J Mol Catal B Enzym 10:471–476

10. Tang Z-X, Qian J-Q, Shi L-E (2007) Characterization of immobilized neutral lipase on chitosan nano-particles. Mater Lett 61:37–40

11. Yamato S, Kawakami N, Shimada K, Ono M, Idei N, Itoh Y (2000) Preparation and characterization of immobilized acid phosphatase used for an enzyme reactor: evaluation in flow-injection analysis and pre-column liquid chromatography. Anal Chim Acta 406:191–199

12. Zhuang P, Butterfield DA (1992) Structural and enzymatic characterizations of papain immobilized onto vinyl alcohol/vinyl butyral copolymer membrane. J Membr Sci 66:247–257

13. Johnston W, Maynard N, Liaw B, Cooney MJ (2006) In situ measurement of activity and mass transfer effects in enzyme immobilized electrodes. Enz Microb Technol 39:131–140

14. Bowers LD, Johnson PR (1981) Characterization of immobilized β-glucuronidase in aqueous and mixed solvent systems. Biochim Biophys Acta Enzymol 661:100–105

15. Constantinides A, Vieth WR, Fernandes PM (1973) Characterization of glucose oxidase immobilized on collagen. Mol Cell Biochem 1(1):127

16. Ford JR, Chambers RP, Cohen W (1973) An active-site titration method for immobilized trypsin. Biochim Biophys Acta Enzymol 309(1):175–180

17. Cooney MJ, Lau C, Martin G, Svoboda V, Minteer SD (2009) Enzyme catalyzed biofuel cells. Energy Environ Sci 1:320–337

18. Invnitski D, Branch B, Atanassov P, Apblett C (2006) Glucose oxidase anode for biofuel cell based on direct electron transfer. Electrochem Commun 8:1204

19. Mikkelsen SR, Lennox RB (1991) Rotating disc electrode characterization of immobilized glucose oxidase. Anal Biochem 195(2):358–363

20. Michaelis L, Menten ML (1913) Die Kinetik der Invertinwerkung. Biochem Z 49:333

21. Lineweaver H, Burk D (1934) The determination of enzyme dissociation constants. J Am Chem Soc 56:658–666

22. Mathebe NGR, Morrin A, Iwuoha EI (2004) Electrochemistry and scanning electron microscopy of polyaniline/peroxidase-based biosensor. Talanta 64:115–120

23. Cooney MJ, Peterman J, Lau C, Minteer SD (2008) Fabrication and characterization of hydrophobically modified chitosan scaffolds. Carbohydr Polym 75(3):428–435

24. Klotzbach T, Minteer S (2007) Improving the microenvironment for enzyme immobilization at electrodes by hydrophobically modifying chitosan and nafion polymers. J Membr Sci 311(1–2):81–88

25. Klotzbach T, Watt M, Ansari Y, Minteer SD (2006) Effects of hydrophobic modification of chitosan and Nafion on transport properties, ion-exchange capacities, and enzyme immobilization. J Membr Sci 282:276–283

26. Martin G, Cooney MJ, Minteer SD, Ross JA, Jameson DA (2009) Fluorescence characterization of polymer chemical microenvironments. Carbohydr Polym. doi:10.1016/j.carbpol.2009.02.021

27. Bard AJ, Faulkner LR (2001) Electrochemical methods, 2nd edn. Wiley, New Jersey

28. Xiaoxia Y, Hong L, Zhenghe X, Weishan L (2009) Electrocatalytic activity of [Ru(bpy)3]$^{2+}$ for hypoxanthine oxidation studied by rotating electrode methods. Bioelectrochemistry 74(2):310–314

29. Svoboda V, Cooney MJ, Liaw BY, Minteer SD, Piles E, Lehnert D, Barton SC, Rincon R, Atanassov P (2008) Standardized characterization of electrocatalytic electrodes. Electroanalysis 20(10):1099–1109

30. Lau C, Minteer SD, Cooney MJ (2010) Development of a chitosan scaffold electrode for fuel cell applications. Electroanalysis 22:793–798

31. Sjoholm KH, Cooney MJ, Minteer SD (2009) Effects of degree of deacetylation on the hydrophobically modification of chitosan. Carbohydr Polym 77(2):420–424

32. Liu CG, Desai KGH, Chen XG, Park HJ (2005) Linolenic acid-modified chitosan for formation of self-assembled nanoparticles. J Agric Food Chem 53(2):437–441

33. Moore CM, Akers NL, Hill AD, Johnson ZC, Minteer SD (2004) Improving the environment for immobilized dehydrogenase enzymes by modifying nafion with tetraalkylammonium bromides. Biomacromolecules 5(4):1241–1247

34. Cooney MJ, Windmeisser M, Liaw BY, Lau C, Klotzbach T, Minteer SD (2007) Design of chitosan gel pore structure: towards enzyme catalyzed flow-through electrodes. J Mater Chem 18:667–674

35. Ma K, Adams MWW (2001) Hydrogenase I and hydrogenase II from Pyrococcus furiosus. Methods Enzymol 331:208–216

INDEX